GRASSES
of East Africa

DINO J. MARTINS

Published by Struik Nature
(an imprint of Penguin Random House South Africa (Pty) Ltd)
Reg. No. 1953/000441/07
The Estuaries No. 4, Oxbow Crescent, Century Avenue, Century City, 7441
PO Box 1144, Cape Town, 8000 South Africa

Visit www.penguinrandomhouse.co.za and join the
Struik Nature Club for updates, news, events and special offers.

First published in 2022

3 5 7 9 10 8 6 4 2

Publisher: Pippa Parker
Managing editor: Roelien Theron
Editor: Natalie Bell
Designer: Dominic Robson
Concept design: Janice Evans
Cartographer: Steve Collins
Proofreader: Emsie du Plessis

Reproduction by Studio Repro
Printed and bound in China by C&C Offset Printing Co., Ltd

MIX
Paper from
responsible sources
FSC
www.fsc.org FSC® C018179

ISBN 978 1 77584 548 5 (Print)
ISBN 978 1 77584 760 1 (ePub)

Back cover: Dino J. Martins and *Sorghum versicolor, Melinis repens* spikelet and aphid
Title page: *Phragmites australis*
Contents page: *Lintonia nutans, Enteropogon macrostachyus*

CONTENTS

INTRODUCTION

This book presents an introduction to the ecology and natural history of East Africa's grasses. It seeks to shine a light on the beauty, diversity and importance of these unassuming plants, and to help specialists and ordinary folk alike to notice and appreciate their value. For the purposes of this book, East Africa comprises Kenya, Tanzania, Uganda, Rwanda and Burundi.

Human beings and grasses have a long and complex history of association. A few million years ago, as Africa became more arid, grasslands replaced the earlier forest habitats. The grasses diversified and spread across the continent to create open habitats that filled with vast herds of grazing animals reliant on the nutritious leaves and stems that made up these new pasturelands. These factors – the opening of the landscape, the ready availability of grass, and the abundance of grazers – created the conditions for our hominid ancestors to evolve from the

primates and to move across the landscape. With their ability to walk on two legs, hominids could advance more efficiently across the grasslands than their animal counterparts and could take advantage of the plentiful prey. Their rich diet, obtained from scavenging and preying on herbivores, would not have been possible without the grasses. The savannas of Africa are where our human lineage was born.

Grasses are one of the world's most familiar plants, yet they are also – paradoxically – one of the most overlooked. When we feel those soft grasses underfoot it is difficult to believe that these humble yet tenacious plants formed the foundation of a food web that supported our early ancestors. In many different ways, grasses are still tied to our success as a species.

The grasses are grouped together in a single plant family, the Poaceae (formerly called the Gramineae). Grasses are monocotyledenous – part of the 'monocots' – along with lilies,

Red Oat Grass is one of the most widespread and nutritious of East Africa's grasses.

East Africa's savanna wildlife populations, such as these wildebeest, depend on healthy, diverse grasslands for their survival; Ngorongoro Crater, Tanzania.

orchids and palms. These plants make up a quarter of all known plant diversity on the planet. Monocots are flowering plants that share a number of characteristics, including seeds (embryos) with a single cotyledon (the first leaf to emerge from inside the seed when it germinates); distinctive parallel venation in the leaves; and flower parts arranged in multiples of three.

Globally, there are over 12,000 described species of grass in some 771 genera; East Africa is home to over 1,000 native grass species, with a number of introduced pasture and fodder species that have also become widespread.

Grasses are found almost everywhere in East Africa. Approaching from the sandy shores of the Indian Ocean, the first plants one encounters are grasses. Here they grow in the wind, salty spray and intense sunshine, helping to provide structure and stability to sand dunes, and shelter for many different species. Moving inland from the coastal areas,

each different habitat has a number of grass species – in the vast drylands and endless bush areas, in woodlands and savannas, in wetlands and swamps, and even in the deep shade of forests, grasses are growing and thriving. Everywhere, grasses make up an important component of the plant community; even on the highest mountains such as Mount Kenya and Mount Kilimanjaro, and in the Ruwenzori Mountains (between Uganda and the Democratic Republic of the Congo), close to the permanent zone of ice and snow, there are grasses clinging to the thin soil or hidden in crevices, where they survive exposure to freezing temperatures.

These beautiful plants are a vital part of the identity and character of East Africa's landscapes. The classic vistas of the East African savannas – so beloved of wildlife photographers and filmmakers the world over – would not be possible without the omnipresent grasses.

ECONOMIC SIGNIFICANCE OF GRASSES

Across the world, grasses are the most economically important group of plants. Besides feeding all the planet's human inhabitants, they also fuel animal agriculture, soften urban landscapes, support the tourism industry, and play a valuable role in the environment and as a natural resource.

Wheat – a grass – is an important cereal crop globally, along with maize (corn) and rice.

Cereal crops

Most of the calories consumed by people across the world on a daily basis come from cereal crops, including maize (corn), rice and wheat; all of these were domesticated from wild grasses by early humans. Rice alone supports at least half of the world's population. Maize, first cultivated in the Americas, has become the staple food for most of the population in sub-Saharan Africa today.

Grasses were the first plants to be domesticated and cultivated by human beings. Wheat and barley were cultivated 11,000 to 12,000 years ago in the Fertile Crescent of the Middle East. Rice was domesticated and cultivated in Asia at least 9,000 years ago, and in Africa around 3,000 years ago. For thousands of years, beer has been brewed from cultivated barley and other important grass crops such as sorghum, millet, oats, rye and sugarcane.

Livestock diet

The livestock sector in East Africa contributes hundreds of millions of dollars to the economy through the provision of protein-rich nutrition to millions of people.

Grasses are the starting point of livestock farming, either as direct grazing or in the form of hay, silage and fodder.

In East Africa, dozens of pastoralist communities have for generations followed rains in search of grazing and sustenance for their herds of cattle, camels, sheep and goats. One of the most satisfying sights to be

Natural pastures are the major feed resource for many indigenous livestock in East Africa, such as this zebu bull grazing in northern Kenya.

had in East Africa is that of healthy herds of livestock or wildlife grazing contentedly on prolific grassland.

Healthy grasslands can support large herds of grazers, and grasses are remarkably adapted to keep growing and sprouting despite repeated grazing.

From the Amazon to Australia, and the United States of America to Israel, a number of grasses that originate in East and southern Africa are important for livestock production, yet there are also invasive species that pose serious management challenges. Landowners use a wide range of strategies to keep these grasses under control, including biological control measures, controlled fires and grazing schemes. Active research in management tactics of invasive species is ongoing.

Urban vegetation

Grasses provide comfortable domestic lawns and grassy land for public spaces and golf courses the world over. It is widely speculated that our fondness of open grassy spaces with occasional trees speaks to our savanna origins in East Africa! Grasses and green spaces in urban areas are essential for mental and physical well-being. Green spaces are shrinking fast in East Africa's cities, yet they offer a comfortable place where families and communities can gather and relax.

Wildlife and ecotourism industry

The grasses in the savannas and grasslands of East Africa form the foundation of a robust wildlife and ecotourism industry. Many lives and livelihoods in this sector are directly or indirectly supported by grasses.

Commercial seed production

An important commercial sector is dedicated to harvesting grass seeds for the establishment of new pastures and the restoration of overgrazed areas. Grass reseeding is an important tool for ecological restoration and returning degraded land to an economically productive state.

Natural material for structural, artisanal, medicinal and spiritual use

In East Africa, and in many other parts of the world, grasses are widely used as thatching and construction material. Artisans fashion grasses into a wide range of utilitarian and decorative items. These plants are also ingredients in traditional medicines and play an integral role in some spiritual practices.

This grass hut was constructed to bring good luck and ward off evil; Tabora, central Tanzania.

The livelihoods of pastoralist communities in East Africa are inextricably linked with grasslands.

STRUCTURE OF A GRASS PLANT

Grass plants have a distinctive structure. A knowledge and understanding of the key parts and features of a grass plant aids identification of grass species in the field.

Grasses are flowering plants, even though their floral parts are not typically showy.

Awns are hairy adaptations on spikelets that help with seed dispersal.

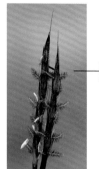

An **inflorescence**, when closely examined, reveals a number of tiny **spikelets** – the flowering and seed-bearing parts.

Culms, or grass stems, are hollow or fibrous, with solid nodes. Culms support the inflorescence and transport water and nutrients between roots and leaves.

Grass florets are part of the spikelet. They remain hidden and their reproductive structures are only visible when the anthers and stigmas are seen dangling from between the lemmas and paleas.

anthers (shed pollen into the wind)

palea (upper bract; encloses floret)

lemma (lower bract; encloses floret)

filament (supports anthers)

stigma (receives pollen on hair-like projections)

ovary (contains ovules that become seeds when fertilised)

Leaf length, width, colour, texture, shape and margins help in identification of grasses. Additionally, identifying how the leaf is attached to the stem, and the membrane at the base of the leaf (ligule), will also assist in distinguishing species.

Grasses grow either from **stolons** above ground (as seen in Kikuyu Grass, right) or **rhizomes** below ground. Grasses have an **adventitious root system**.

Inflorescence types

The flowering parts of a grass form part of the inflorescence, which can take different shapes, including a spike, raceme or panicle, or a variation of these.

Here are some of the most common types of inflorescence to be found among the grasses.

Spikes are a form of inflorescence where all the spikelets are closely packed together along the main axis. Grasses with this kind of structure include Buffel Grass (*Cenchrus ciliaris*) and Riparian Wire Grass (*Pennisetum riparium*). Spiked inflorescences are seen in most of the wire grasses (*Pennisetum* spp.).

Racemes consist of a series of short branches from the central stem (axis), typically angled to one side. Grasses with this form of inflorescence include Hairy Signal Grass (*Brachiaria lachnantha*), Palisade Signal Grass (*Brachiaria brizantha*) and Ditch Millet (*Paspalum scrobiculatum*).

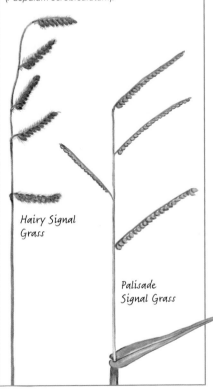

Buffel Grass

Riparian Wire Grass

Hairy Signal Grass

Palisade Signal Grass

Panicles are generally more spread out than the other inflorescence types. A number of branches extend from the central stem and point in different directions, giving an overall symmetrical appearance. Grasses with this inflorescence form include most of the love grasses (*Eragrostis* spp.), dropseed grasses (*Sporobolus* spp.) and Guinea grasses (*Panicum* spp.).

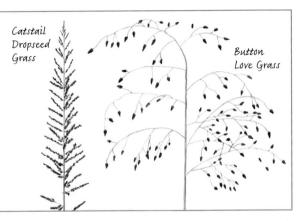

Catstail Dropseed Grass

Button Love Grass

ECOLOGY OF GRASSES

Grasses may be the most successful plant family on the planet, making up the bulk of the biomass in many different habitats. Plants of the grass family, and the other forms of life they support, are dramatically and beautifully evident in the many grassland and savanna national parks in East Africa. Massive herds of grazers, such as wildebeest or gnu, can be seen moving through the Mara–Serengeti ecosystem every year, preferentially feeding on grasses when they are available, while thousands of elephant, the largest land mammal, also partake of the grasses.

These grasses are tough plants with multiple adaptations that allow them to disperse, establish in and colonise new ground, persist through periods of drought and flood, and survive competition, intense grazing and trampling by herbivores. Grasslands make up over 40 per cent of all the known terrestrial habitats on Earth. These areas of high primary productivity are extremely efficient at converting sunlight and water into sugars through photosynthesis, creating the basis of the food web that supports so much other life.

These plants have been with us for a long time, as evidenced by grass remains found in the fossilised dung of some dinosaurs. Grasslands as a dominant habitat spread during the Miocene period, between 23 and 5.3 million years ago. This epoch also saw the development and evolution of the apes. In the early Miocene, C3 grasses spread and were successful, and in the later Miocene,

CARBON & NUTRIENT CYCLE ON THE AFRICAN SAVANNA

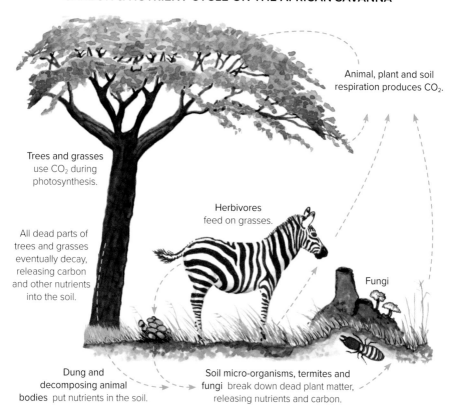

Animal, plant and soil respiration produces CO_2.

Trees and grasses use CO_2 during photosynthesis.

All dead parts of trees and grasses eventually decay, releasing carbon and other nutrients into the soil.

Herbivores feed on grasses.

Fungi

Dung and decomposing animal bodies put nutrients in the soil.

Soil micro-organisms, termites and fungi break down dead plant matter, releasing nutrients and carbon.

The magnificent megaherbivores of Africa are a key part of grassland ecology; Laikipia, Kenya.

C4 grasses emerged and established dominance. C4 grasses are primarily tropical species that developed a slightly different form of photosynthesis that is more efficient at higher temperatures and simultaneously saves water, while C3 grasses have a more typical form of photosynthesis. Maize and sugarcane, both important crops, are C4 grasses.

Grasses help fix carbon from the atmosphere into their soils, increasing the Earth's carbon content through their extensive root systems. As habitats of high productivity, grasslands have helped form and maintain robust, fertile soils. Many high-production croplands are located in areas that were formerly grasslands.

Life cycles and seed dispersal

Most grasses are either annuals or perennials, but some species can adopt either lifestyle, depending on the local conditions.

One of the most interesting adaptations that grasses have is the ability to disperse their seeds. Many grass species have spikelets that bear awns, which are often armed with hooks, barbs, hairs or spines enabling them to attach to materials in the environment such as fur, feathers or clothing – as anyone who has walked through a field of Short or Tall Spear Grass can attest! The lemmas, which are part of the spikelet, often have similar structures to the awns. This ability of grass seeds to 'hold on' enables them to be carried long distances, across land and sea, persisting on their unwitting dispersers for many months in some cases. Other grasses can disperse in the bellies of herbivores; they are carried safely to new areas where they are deposited in piles of dung.

Even more remarkably, some of the grasses with long-awned seeds, such as the Red Oat Grass (*Themeda triandra*) and various thatching grasses (*Hyparrhenia* spp.), can actually bury themselves in the soil. Their awns are hygroscopic, twisting as the humidity varies in a spiralling, corkscrew motion that pushes the seed into the soil, where it can survive drought, fire and the unwanted attention of rodents or birds. Once the rains return, the seed germinates and a new plant is established. This is just one example of how resilient grasses are, and why they can establish as important pioneer species on bare ground where other plant species might struggle to survive.

Succession and pioneer species

Grasses play an important role in the succession of vegetation in natural habitats and cultivated areas. In many parts of East Africa, land is cultivated seasonally. A piece of ground is cleared of natural vegetation, crops are grown there for a few seasons, then the land is left fallow, to rest and recover, before being brought into cultivation again. Several grasses rapidly establish on recently cultivated or bare ground after crops have been harvested. These 'pioneer species', as they are known, are the first plants to emerge, helping to stabilise the soil, provide ground cover and attract insects, birds, rodents and other creatures. This biological activity creates an environment where other species can grow and thrive.

GRASSES AND GRAZERS

One of the most remarkable things about grasses is the diversity and quantity of animals that feed on them. Grasses are able to support vast numbers of grazers and they survive intense grazing. They do this primarily by keeping the growing point hidden at the base of the plant. This allows new shoots to keep sprouting from the individual plant so long as the conditions are favourable for growth. Across East Africa, the main factor that influences grass growth and availability is moisture/rainfall. All East African grasslands and savannas have an element of seasonality to them, with distinct wet and dry seasons.

Grasses are widely eaten by a large number of herbivores, with dozens of other mammal species depending on them for a significant part of their diet. However, overall, grasses are difficult to digest, as most of the energy in the grass is tied up in tough fibrous tissue in the form of cellulose. Animals that feed on grasses have developed various ways to digest them. Adaptations by grazers include the way in which they feed, the physical form of their mouths, lips and teeth, the structure of their digestive system, and how selective they are when feeding.

Grazers and browsers

Herbivores that feed on plants can be broadly divided into grazers and browsers. Grazers feed mostly on grasses while browsers eat a mixed diet, with grass forming a variable component of their food intake. Grazers have distinctive teeth, with high-crowned molars that are adapted to eating grass, sustaining more grinding and wear and tear; this is known as hypsodont dentition. Grass stems and leaves are covered in silica, which can cause great wear on teeth over time. Grazers, such as zebra, use their incisors to nip off grasses, then their molars to chew them and break them into smaller pieces before they swallow.

The digestive systems of herbivores also show adaptations to their diet. All herbivores use varied forms of fermentation to help digest their food. Foregut fermenters are known as ruminants – these animals have a four-chambered stomach. They feed and fill the first chambers (namely the rumen and reticulum),

Grasses, like this tall Thatching Grass, provide both grazing and cover for mammals, such as the eland seen here, as well as many birds and insects; Nairobi National Park, Kenya.

The wide, square lip of the white rhinoceros is an adaptation to grazing on short, dense grass.

then they regurgitate and chew the food more thoroughly while resting, passing it into the next chambers (the omasum and abomasum) where digestion proceeds. Ruminants include impala, waterbuck, bushbuck, dik-dik, buffalo, cattle and other species such as camels and giraffes. 'Chewing the cud' is a typical behaviour of the ruminants.

Hindgut fermenters, such as zebra and elephant, use another system for digesting plants. Zebra feed almost exclusively on grasses and are able to make use of grass that is not easily consumed by other species, while elephants eat grasses when they are available. Hindgut fermenters are bulk feeders that pack their single-chambered stomachs full of food; digestion happens slowly, in an enlarged pouch called the caecum, with the assistance of specialised microbes. Specially adapted microbes and fungi have evolved in the herbivores and are essential for ruminants and hindgut fermenters.

Resource partitioning and facilitation
Two important ecological features of grazers are resource partitioning and facilitation. When many different animals share a common resource such as grass, there is selection to evolve ways to reduce competition between them – this is known as resource partitioning.

While there are many different grazers in East Africa, close observation shows that even in a small area, they will feed in slightly different spots, at different times, and even on different individual species of grass. Facilitation refers to a type of positive interaction between species that benefits both participants or does not harm either of them. For example, as buffalo and zebra feed on coarse, tall grasses, they help open up the grassland for a variety of species such as cattle, gazelles and wildebeest that need access to shorter, more nutritious grasses.

Overgrazing and undergrazing
Grasses and grasslands can be overgrazed or undergrazed. Overgrazing results in a loss of species diversity and an increase in areas of bare ground, which leads to soil erosion and ultimately a collapse of the robust grassland nutrient cycle. Undergrazing, which is much less common, produces a grassland with an abundance of moribund grass. This dry, standing grass is of little nutritional value from a grazing perspective; more woody plant species will eventually establish here as a natural succession. Grazers need grasses, and grasses need grazers. Their interactions maintain the dynamic system that supports healthy grasslands and herbivores – a system in which nutrients cycle between grass, grazers and soils.

GRASSES AND ARTHROPODS

The main group of herbivores that feed on grasses and inhabit grasslands are the insects and some other arthropods. A single acre of grassland contains many insect species, including grasshoppers and locusts, caterpillars of various moths and butterflies, and many different flies, beetles, aphids, plant bugs, scale insects and mites. Among the most noticeable of grass-feeding insects are the grasshoppers and locusts. These are seasonally present in the grasslands of East Africa, with their eggs hatching when the rains return and the grasses start growing again; insect numbers can add up significantly at these times. It is estimated that there are 10–20 tonnes of grasshoppers per square kilometre in the Mara–Serengeti ecosystem and 5–10 tonnes on the plains of Laikipia. Even in the arid areas of Northern Kenya–Uganda, there are 0.5–0.75 tonnes of grasshoppers per square kilometre. This far exceeds the number of herbivorous mammals in all these sites.

An important group of insects that is closely tied to grassland ecology is the harvester ants (*Messor* spp.). From the shores of the Mediterranean all the way through East Africa and down to the Cape of Good Hope, harvester ants play an important role in collecting, dispersing and storing grass seeds. Several thousand years ago, biblical legend King Solomon wisely and accurately observed these very same ants, and counselled his audience: 'Go to the ant, thou sluggard! Consider her ways and be wise!' Harvester termites comprise another group of insects that feed on grasses, and by pruning the plants create lush lawns in wetter areas.

A number of butterfly and moth species use various grasses as host plants for their caterpillars. Many species of 'browns', in the family Nymphalidae, have larvae that are grass feeders, including the evening browns (*Gnophodes* and *Melanitis* spp.). The bush brown (*Bicyclus safitza*) has become important for science, as studies of this butterfly have given insights into the genetics and evolution of many features, including ageing, body size, wing patterns and behaviour. Dozens of different flies and beetles also make use of grasses, by feeding on the leaves and as larvae inside the stems – these larvae are known as stemborers.

Ticks and mites (order Acari) are another important group of arthropods that live in grasslands. Ticks are often found clinging to various grass species, from where they can attach to passing hosts. As ticks are vectors of diseases, they can have a serious impact on cattle, livestock and people if their numbers are not kept in check. In healthy grasslands with a high diversity of species this job falls to the tick predators – spiders, ants, lizards and birds – and a range of naturally occurring fungi and bacteria that infect these parasites. The high volume and variety of mammals that move through the grasslands also help to reduce the size of the tick population – by 'mopping up' ticks with their coats as they rub against the grasses and transporting them to other locations. When livestock are treated in dips or sprayed, this also helps to reduce the tick population. To preserve the resilience of grasslands, and the health of the animals that live and feed on the grasses, it is important for these habitats to have a diversity of predators and parasites that feed on ticks. An unfortunate consequence of overgrazing is that it may weaken a grassland's ecosystem – research data have shown that overgrazed areas often have a higher number and density of ticks.

Ticks, their hosts and grasses are involved in a number of complex interactions that are currently being investigated by scientists.

Harvester ants (*Messor* spp.) are important dispersers of grass seeds.

GRASSES AND FUNGI

Grasses and mushrooms colonise abandoned livestock enclosures.

Fungi occur almost everywhere and are very diverse. However, they have been poorly studied and scientists are only now beginning to appreciate the important roles that they play ecologically and economically. Fungi and grasses have many diverse interactions, and fungi can help or harm grasses. Many fungi live within the leaf or stem tissues of a grass, enjoying a mutualistic relationship with their host. Known as endophytes, these fungi play an important role in limiting grazing by herbivores because they produce compounds that give grasses an unpleasant taste. Mycorrhizal fungi live in the roots of plants, including grasses, and are important for nutrient absorption; they even help nutrients move between plants.

Some fungi are pathogenic, however, and cause diseases in grasses. Rusts are one group that typically infects leaves and stems, causing colourful patches to appear. Smut fungi tend to be more powdery and infest the flowering parts of grasses. Ergots grow on some grasses' spikelets, producing swollen fruiting bodies. Rusts and smuts in grasses can have several different stages and even more than one host, moving between different species. The impact of fungal diseases in grasses has been long evident, from famines described in ancient texts, to the more recent emergence of a wheat rust, UG99, that is spreading rapidly across the world.

A smut fungus is clearly visible on Buffel Grass.

Many different rust fungi grow on grass leaves.

These Coastal Turf Grass spikelets are infested with fungi.

GRASSLAND CONSERVATION AND MANAGEMENT

Many human activities impact grasslands, but the two main culprits are the conversion of grasslands for growing crops, and overgrazing as a consequence of increases in livestock densities. Both of these activities are occurring on a global scale.

A recent analysis of the biomass of terrestrial ecosystems found that, for terrestrial mammals, 96 per cent of the standing biomass is made up of humans and livestock, with just four per cent being wildlife. The grasslands and savannas of East Africa have some of the highest mammal diversity recorded. Yet across many parts of East Africa, wildlife has been declining steeply; in Kenya it has reduced by 60–80 per cent over the past few decades. Despite these alarming trends, East Africa still contains some of the world's most spectacular grasslands and wildlife populations.

East Africa's seasonal rainfall

This region's grasslands and savannas are dependent on seasonal rainfall that comes from a monsoon-based system around the Intertropical Convergence Zone (ITCZ). This is a band of weather activity encircling the planet in the region of the equator. It is a dynamic system that moves across the globe in a north–south pattern following the Earth's tilt and the Sun. This system produces the distinct seasons seen in the northern and southern hemispheres, as well as the wet and dry seasons. The first rainy season begins as the ITCZ moves north; winds from the southeast monsoon bring the 'long rains' to the land from clouds formed over the Indian Ocean. This rainy season starts in March in Tanzania, and moves north into Kenya and Uganda through April and May. The ITCZ reaches its northern limit in July (mid-summer for the northern hemisphere), then it starts moving south. There is a short dry season in most of the region between July and October. Driven by the southwest monsoons, a second rainy season, known as the 'short rains', is experienced in November and December. The main dry season runs from January to March, when the highest temperatures are seen here.

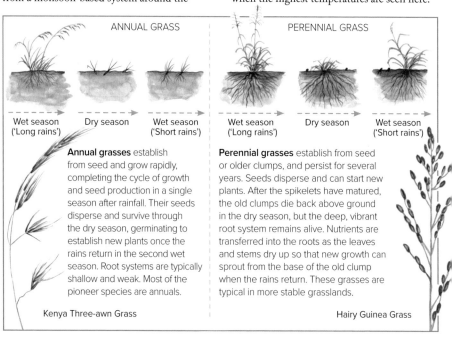

ANNUAL GRASS

Wet season ('Long rains') — Dry season — Wet season ('Short rains')

Annual grasses establish from seed and grow rapidly, completing the cycle of growth and seed production in a single season after rainfall. Their seeds disperse and survive through the dry season, germinating to establish new plants once the rains return in the second wet season. Root systems are typically shallow and weak. Most of the pioneer species are annuals.

Kenya Three-awn Grass

PERENNIAL GRASS

Wet season ('Long rains') — Dry season — Wet season ('Short rains')

Perennial grasses establish from seed or older clumps, and persist for several years. Seeds disperse and can start new plants. After the spikelets have matured, the old clumps die back above ground in the dry season, but the deep, vibrant root system remains alive. Nutrients are transferred into the roots as the leaves and stems dry up so that new growth can sprout from the base of the old clump when the rains return. These grasses are typical in more stable grasslands.

Hairy Guinea Grass

This bimodal rainfall system is distinct from that of southern and western Africa, where there is typically just one wet and one dry season each year. As the rainfall in East Africa is spread out over two rainy seasons, grasslands are more widespread as they are better adapted to these drier, seasonally fluctuating conditions. Grasses here are adapted to take advantage of localised rainfall, and they grow rapidly, producing flowers and seeds after the rains. Even just a few days of rain can transform a landscape from dry and dusty to green and flourishing – an indicator of the grasses' ability to respond and grow new leaves rapidly when conditions are favourable.

Climate change is disrupting these long-established weather patterns; longer periods without rainfall (droughts) are being recorded as well as increases in the time between rainy seasons. Storms, when they do occur, are heavier and more intense.

Livestock density

Livestock are an important part of the culture and economy in this region. For thousands of years pastoralist communities have moved across the landscape with herds of cattle, camels, sheep and goats. The number, distribution and density of livestock have been changing in recent times. People move around less than in previous generations, and more livestock are surviving as a result of better vaccination programmes and veterinary care. Consequently, many grassland areas are experiencing overgrazing. Even though grasses are adapted to being grazed, when they are too heavily grazed and trampled, they eventually disappear, leaving the soil bare and exposed to erosion from wind, water and the cumulative effect of millions of hooves. Soil erosion and land degradation are a very serious issue facing many parts of East Africa.

The importance of fire

Historically, one of the characteristic events in the grasslands of East Africa has been fire. Fires occur naturally, through lightning strikes, or deliberately, when they are started by people. Small local fires and other disturbances surprisingly increase grassland diversity. But large fires and the repeated deliberate burning of grasslands under the wrong conditions drastically reduce grassland diversity and resilience.

Grasses are one of the most important groups of plants we have for solving the many challenges facing our planet. They can be used for restoring degraded land, controlling soil erosion, reducing atmospheric carbon by fixing it in the soil, and keeping nutrients cycling in the food web. We need to understand, appreciate, manage and protect grasses better to preserve species and land for future generations.

Water towers of East Africa

Across East Africa, the mountains and highland areas play an important role as 'water towers' for a thirsty region. These areas of land at higher altitude are covered with natural vegetation that captures rain and releases it slowly into streams and rivers as part of the watershed. Millions of people living in the major cities and surrounding areas in the region depend on these water resources. One particular grass species, the African Mountain Bamboo (*Yushania alpina*) – seen as pale green patches in the photograph below – is a crucial part of this system. It grows in a distinctive zone on all the large mountains and montane regions of East Africa, ranging from 2,700–3,400 metres in altitude. Within this bamboo zone, it is the dominant plant that helps to trap rainfall, feeding it into the many small streams that flow down the slopes.

HABITATS OF EAST AFRICA

East Africa is one of the most diverse regions on the planet – travellers can move from ice-covered rocky mountains to hot, dry deserts in just a few hours. The varying climate is driven by the interesting geography, which includes the Great Rift and Albertine Rift valleys that run north–south through the region.

Rainfall is the key driver of different habitats in this region. The many habitats here reflect the diverse vegetation zones and characteristics of a particular geographic area. Within even a single habitat there can be a wide range of land uses and vegetation types, especially where human activities, such as agriculture, grazing of livestock and use of fire, and other impacts are evident. Many areas are now under cultivation for agriculture, and in some cases have been farmed or under crops for almost a century.

Grasses are widely distributed across all of East Africa's terrestrial habitats. They also grow along the edges of freshwater and coastal zones. Many can be found within several different habitats, and at a range of altitudes, while others are more localised in their distribution.

Below are some of the major habitats in East Africa, with examples of the typical grasses they contain.

Wetlands

Throughout East Africa, wetlands form where local conditions allow for water to accumulate from rain, runoff or underground sources. These wetland habitats include marshes, swamps and seasonally waterlogged pans where soils have a layer of rock beneath them. A number of grasses are adapted to wetlands and play an important ecological role within them by providing natural filtration, stability and habitat for many different creatures.

Forest

Forests occur where there is sufficient rainfall, typically over 1,850 millimetres per year. This habitat is not typically rich in grasses, which tend only to grow on the forest floor and along roads, paths, tracks and trails. Many of East Africa's forests have been cleared to make way for agriculture, with the remaining forests also under threat; conservation is focused on the role they play in protecting biodiversity and serving as essential water towers. The most important grass species in montane forest zones is bamboo, and on many high mountains, there is a distinct bamboo zone.

Forest habitat; Kakamega Forest, western Kenya

MAJOR VEGETATION ZONES IN EAST AFRICA

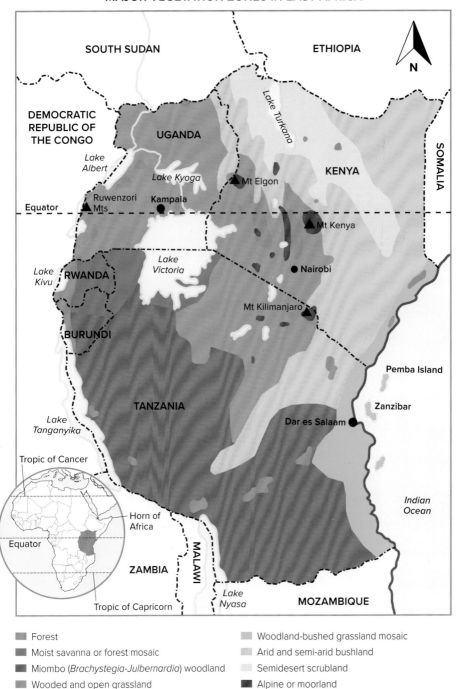

Legend:

- ■ Forest
- ■ Moist savanna or forest mosaic
- ■ Miombo (*Brachystegia-Julbernardia*) woodland
- ■ Wooded and open grassland
- ■ Woodland-bushed grassland mosaic
- ■ Arid and semi-arid bushland
- ■ Semidesert scrubland
- ■ Alpine or moorland

Moist savanna; western Tanzania

Moist savanna or forest mosaic

Moist savanna or forest mosaic occurs primarily south of Lake Victoria and in the western regions of Tanzania as well as throughout Rwanda and Burundi. It is a highly seasonal, typically humid habitat with patches of forest and woodland where tall grasses grow in glades. Grasses tend to be present all year round, even during the dry season. Trees tend to be large and at varying densities, and include large acacias, tall stands of palms and species of *Albizia*.

Miombo (*Brachystegia-Julbernardia*) woodland

Miombo woodland is a classic habitat of central and southern Tanzania, extending further south towards southern Africa. Miombo woodland consists of a unique assemblage of trees and shrubs that are partly deciduous. Many of the trees flower synchronously at the beginning of the long rains in March and this is followed by a period of intense plant growth and consequent ripening of seeds. Much of this habitat remains intact and is still relatively little explored in terms of biodiversity. This is a very seasonal habitat with many different grasses that appear after rains.

Miombo woodland; central Tanzania

Wooded and open grassland

Wooded and open grassland is a widespread habitat type that stretches from northern Tanzania into Kenya's Rift Valley, over the central and western plateaus of the region, into Uganda. Quintessential vegetation here consists of mixed grasses and herbs that serve as ground cover, together with a scattering of trees. There are some unique microhabitats here, including grassland with acacias that have evolved to house ants in swollen thorns. Known as 'whistling thorns', these trees tend

Wooded and open grasslands are a feature of many of East Africa's national parks; Tarangire, Tanzania.

to grow on seasonally wet black-cotton clay soils, together with a number of specialised grasses that grow only on these soils. This habitat is undergoing widespread change as a result of overgrazing, clearing for charcoal and cultivation. This is one of the most diverse habitats for grasses, with dozens of different grass species to be found in a typical open grassland.

Woodland-bushed grassland mosaic

Woodland-bushed grassland mosaic is a habitat of the coastal regions of Kenya and Tanzania. It occurs where the landscape is still close enough to the ocean to receive seasonal rainfall more reliably than further inland, and incorporates coastal coral rag thickets – which grow on rocks derived from corals – and evergreen bush. This habitat is botanically rich with several endemics, including a significant number of coastal or near-coastal endemics. The vegetation can be difficult to negotiate as it is very thick and tangled. Some areas have been cleared to make way for coconut, cashew and mango plantations and small-scale subsistence farming. This is where most of the

shifting cultivation – using slash-and-burn with fallow cycles – is practised. As the soil quality of cultivated land rapidly declines, the land is abandoned after a few years, allowing for grasses and other plants to colonise it. Many different grasses, especially perennial species, occur in this habitat.

The dense woodland-bushed grassland mosaic is seasonally filled with grasses; Jilore, Kenya.

Habitats of East Africa ■ 21

Arid and semi-arid bushland

Arid and semi-arid bushland occurs in the interior regions from the northern part of Tanzania to parts of northeastern Uganda and northwestern Kenya. These dryland areas are often called rangelands in reference to the wide-ranging livestock that pass through in search of pasture. These rangelands consist of a mixture of low-growing thorny shrubs – mainly acacias – as well as a diverse mix of *Commiphora*, *Boswellia* and members of the caper family (Capparaceae). A fairly widespread habitat, it is threatened by overgrazing and large-scale charcoal production.

Historically these areas were considered xeric (drought-tolerant) grasslands, but many decades of overgrazing have transformed them from grasslands to less diverse areas with thorny vegetation. Grasses are present here after rains and consist mainly of fast-growing annuals.

Semidesert scrubland

Semidesert scrubland is one of the most arid habitats of East Africa, with rainfall under 200 millimetres per year. These scrubland areas occur in northern Kenya where there

Semidesert scrubland is one of the harshest environments in the region; Turkana, Kenya.

Arid and semi-arid bushland occurs in much of northern Kenya and Uganda; Ileret, Kenya.

Alpine/moorland habitats have a unique grass community dominated by tussock grasses; Mount Elgon, Kenya–Uganda border.

are extremely hot, arid conditions with many years where little or no rain falls.

As with arid and semi-arid bushland, these areas fall within the rangelands, where grasses may appear seasonally, following rains. The vegetation is mainly low-growing, spiny *Indigofera* and thinly scattered acacias that are adapted to arid conditions.

Grasses are an important seasonal feature of vegetation in semidesert scrublands; even if they do not appear every year, the grasses are tolerant of the extreme heat and aridity in this zone. This habitat is undergoing some changes due to ongoing overgrazing and charcoal production.

Alpine or moorland

Alpine/moorland is a distinctive habitat that occurs on the tops of high mountains and mountain ranges in East Africa, typically above 3,000 metres. It is also known as the Afro-alpine zone and is primarily made up of montane grasslands with tussock grasses as the dominant species, together with giant, perennial, herbaceous and woody plant species. The unique flora here is adapted to the wide temperature range (freezing and thawing daily) and intense sunlight. Many specialised grasses occur in this zone, including species of *Festuca* and the highest-growing grass in East Africa, the Montane Pentameris (*Pentameris minor*).

HOW TO USE THIS BOOK

The grass species described here include those that are frequently encountered, ecologically important, or noticeable. By studying the photographs and illustrations carefully, you will become more adept at recognising many of the region's grasses.

2 · **1** · **7**

Plume Grass

Tetrapogon roxburghiana
(= *Chloris roxburghiana*)

Height 50–150cm

Other name: Feather-duster Grass

Description: This striking perennial grass is one of the most recognisable and beautiful of the East African grasses. It grows in dense stands with leaves tightly pressed together and overlapping. Leaves are 10–45cm long. The distinctive inflorescence is bright purple to pink, fading to pale pink and light green, and eventually to pale and creamy straw coloured. The spikelets are densely clustered on the inflorescence and detach in numbers when the grass is maturing; they have 3–4 wispy awns of equal length.

3

Distribution & Habitat: A widely distributed species across East and southern Africa, the Greater Horn of Africa and parts of India. It is found in bushland and woodland, and thrives both in glades in these habitats as well as in more open areas, including on disturbed ground.

4

Ecology & Uses: An important species for grazing when young by livestock and wildlife in dryland areas. This is a robust and fast-growing species that regenerates quickly, and because it is a highly visible species, it is often targeted for grazing by herders. It is also used as an ornamental plant, both in gardens and flower arrangements.

5

This is one of East Africa's most distinctive grasses, symbolic of the many wild and open drylands in the region.

Younger inflorescences are a darker shade of purple, fading to a bright pinkish colour.

2–3mm

spikelet

Plu

6

...air to the landscape following good rains.

Plume Grass ■ 217

Features of species descriptions

1 Common name Gives English common name where possible, as it may be confusing for non-scientists to decipher scientific names. Common names can vary depending on where they are used, and some of the grasses have different common names in other regions. The common names presented are the English ones most widely used in East Africa.

2 Scientific name Gives the species name. Where necessary, older scientific names are provided in parentheses.

3 Description Presents the main features of the grass, including its size, colour, shape and any other distinctive characteristics that may aid identification. These descriptions are meant to be broad guidelines only; many grasses are highly variable, depending on the local conditions where they grow. This book focuses on the inflorescence and spikelet for identification, as these are the most easily observed features. Observe them closely as they will help you to tell the grasses apart.

4 Distribution & Habitat Indicates where a grass is found and in which habitat it occurs. An altitude range indicates where the grass grows best and, where relevant, the kinds of soil in which it grows.

5 Ecology & Uses Gives general ecological and natural history information and describes a grass's significance for grazing, or which mammals, birds or insects make use of it. For grasses that are used in fodder, or in other practical ways, these details are also provided.

6 Spikelet measurement Provides length of spikelet in millimetres and excludes the variable length of the awns.

7 Illustrations and labels Highlight notable features. Spikelet illustrations have been reproduced to make the small elements visible to the reader – they are not to scale.

Tips for studying grasses

The guidelines below should enhance your study of these beautiful and interesting plants.

- Explore different habitats and spend time observing grasses. Many species are seasonal, so it is important to begin your exploration following rains and when grasses are in flower. Take time to look, listen, watch, observe and record.
- Look closely at the inflorescence and spikelet. Many grasses may look similar from a distance, but it is through close observation that you will learn more about the structure and diversity of grasses. A loupe (hand lens) or magnifying glass will help to focus on the detail.
- Keep records and make notes about the grasses you encounter: write down what you see and note the time, date, season, soils, location, uses and other interesting features. Take photographs if you can.
- Grasses can be pressed and dried easily in a notebook or exercise book. Once dry, they can be labelled and used for reference. Keep these in a cool, dry place, away from insects and damp. If stored well, these specimens can last many years.
- Visit a herbarium and speak to a botanist. The East Africa Herbarium at the Nairobi National Museum has an extensive collection of grasses from the region. You can also look at many herbarium specimens online.

Grasses offer a relatable and accessible way to learn about our environment.

A notebook can be used to collect and press grasses for identification.

Herbarium specimens of grasses are useful for research and study.

Flowering Kikuyu Grass. This species has been widely planted all over the world; the anthers that shed pollen are clearly visible as yellow structures atop the slender silvery filaments.

Highland Andropogon

Andropogon lima   Height 50–200cm

Description: A tall, elegant perennial grass that forms dense clumps. It grows from a hidden rhizome in a tufted manner, forming attractive, slightly irregular stands. Leaves are flat, 1–20cm long. The inflorescence is typically made up of 2 short, finger-like racemes held together fairly closely on the end of a long, straight stalk. Racemes are straight, often with one longer than the other; they have a bright reddish-purple colour when young, becoming more untidy and paler as they age and ripen. The spikelets bear awns 3–5mm long. The ageing leaves persist into the dry season.

Distribution & Habitat: A fairly common and widespread grass throughout East Africa. It grows in the highland areas in upland and montane grasslands, and at the edges of roads and tracks through highland forest. Found at a wide range of altitudes from 2,000–4,000m above sea level.

Ecology & Uses: Grazed by both livestock and wildlife in highland areas. Young grasses of this species are eaten by cattle, and form an important part of the diet of moorland species such as duiker and rodents. The persistent clumps provide shelter for various insects, birds and other plants.

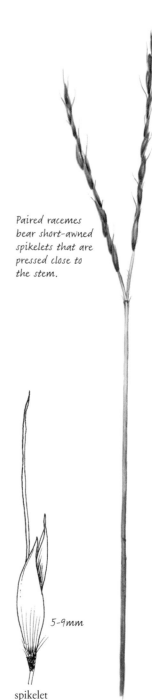

Paired racemes bear short-awned spikelets that are pressed close to the stem.

Tufts rise up from a concealed rhizome.

5-9mm

spikelet

Dense clumps form, with both fresh and older leaves.

Two paired racemes atop a stalk

Clumps of this grass persist into the dry season and provide shelter for many creatures.

Stab Grass

Andropogon schirensis   Height 40–200cm

Description: A tall, wispy perennial grass that forms dense clumps. It often stands elegantly above other grasses. Leaves are flat, 9–70cm long. Old leaves and stems clustered at the base of the plant weather into loose fibres. The inflorescence is typically made up of 2 short spikes held in an open V shape on the end of a long, straight stalk. These spikes are straight and a greenish-cream colour when young, becoming curved and pale straw coloured as they age and ripen. The awns are 2–5cm long and become twisted as the spikelet ripens.

Distribution & Habitat: A fairly common and widespread grass throughout East Africa. It grows in open and tall grassland, woodland and glades, and on rocky escarpments or hillsides. Found at a wide range of altitudes from 250–3,000m above sea level.

Ecology & Uses: Grazed moderately by both livestock and wildlife. In some habitats, including the high plains of Kenya, it is grazed by the endangered Jackson's hartebeest. When other grasses are depleted or scarce, herbivores will feed on this species, especially when it is young and in new leaf. It is locally cut for fodder in a few areas.

Spikes fade with age as they ripen.

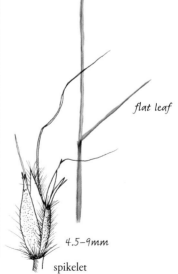

Tall stems bear paired racemes that are purple and greenish when young.

flat leaf

4.5–9mm
spikelet

Stab Grass grows in open grassland.

Mountain Anthoxanthum

Anthoxanthum nivale   Height 20–70cm

Description: A striking, erect perennial grass that grows from a hidden rhizome. Stems are short, rarely taller than knee height. Lower stems are clothed in old, papery leaves, while the fresh upper leaves cling to the stem and point upwards. Leaves, 5–15cm long, are stiff and slightly rough to the touch, with a blunt tip. The inflorescence is a compact, spike-like panicle, with tightly pressed spikelets arranged against the stem. It is dark purple to bright green when young, fading to a pale husk once the seeds have been shed.

Distribution & Habitat: A grass of the highland and montane zones. It grows in damp, moist soil, mostly along the edges of paths, in moorland and the Afro-alpine zone on Mounts Kenya, Elgon and Kilimanjaro. It is also found along the edges of streams and high-altitude bogs. Mountain Anthoxanthum typically grows from 2,400–4,800m above sea level.

Ecology & Uses: Often grows interspersed with clumps of other high-altitude, tussock-forming species of the genus *Festuca*. Grazed by high-altitude mammals such as the common duiker. The seeds are eaten by moorland rodents and birds, including Mount Kenya's endemic Jackson's francolin.

Jackson's francolin is known to eat the seeds of this grass.

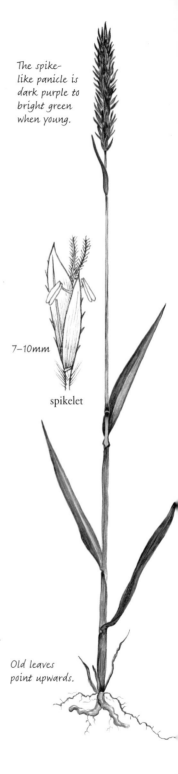

The spike-like panicle is dark purple to bright green when young.

7–10mm

spikelet

Old leaves point upwards.

Young, dark purple inflorescence

Old, faded inflorescence

The purple and green inflorescences are conspicuous in moorlands.

Elegant Three-awn Grass

Aristida adoensis Height 20–60cm

Description: A beautiful medium-sized perennial grass. Narrow leaves, 5–25cm long, have inward-rolled margins and are held close to the stems. The inflorescence is a spike, flecked green with reddish markings when young and typically held slightly curved. As the grass matures, it develops an untidy spiky appearance and fades to light brown. The mature plants form narrow, sparse clumps with distinctive feathery seedheads that wave in the softest breeze.

Distribution & Habitat: A common and widespread grass in bushland and open grassland regions. It grows at mid-altitudes from 1,300–2,400m above sea level.

Ecology & Uses: An important grazing species in drylands and semi-arid regions, where it is grazed by livestock and wildlife. It is grazed less as the plant matures, because the leaves become harder to eat and digest. A key pioneer species by virtue of its tenacious seeds, which embed themselves in clothing, fur and wool, thereby facilitatating wide dispersal by humans and animals.

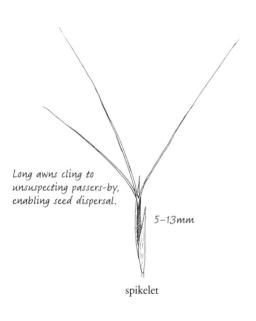

The inflorescence is gently curved; flecked green with reddish markings.

Aged inflorescences fade to light brown and look tatty.

Long awns cling to unsuspecting passers-by, enabling seed dispersal.

5–13mm

spikelet

Young inflorescences are green and slightly curved.

Seedheads become wispy and beige as they age.

Elegant Three-awn Grass thrives in areas of open grassland.

Kenya Three-awn Grass

Aristida kenyensis Height 10–60cm

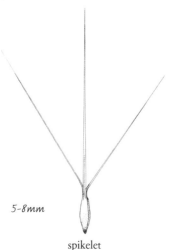

Description: An annual grass that stands erect or slightly arched, growing from a tight cluster of leaves and stems. Leaves, 5–20cm long, are narrow, fairly short and often folded, especially in drier areas. Young spikelets are green with reddish-brown or purple markings, and wave attractively in a gentle breeze. Older spikelets turn pale, with spiky awns protruding from the seeds; they are very brittle and readily detach. The seedhead is typically curved and bends over in young plants.

Spiky awns protrude from the seeds.

Distribution & Habitat: A widely distributed grass species in drylands across East Africa, found mainly in northern Kenya and northern Uganda, and in the dryland areas of northern and central Tanzania. Typically grows on dry, friable soils in acacia bushland or woodland. This species can be common on roadside verges and may form stands on stony ground. It grows from 600–2,130m above sea level.

Ecology & Uses: A fast-growing species that establishes rapidly after rains. Young leaves are grazed by livestock, with grazing reducing as the plant matures. It is an important part of the diet of dryland grazers, including the endangered Grevy's zebra. Mature spikelets readily shed seeds, which embed themselves in clothing or fur, facilitating dispersal.

5-8mm

spikelet

Freshly sprouted leaves and inflorescences

Dry, ripening inflorescence

A young greater kudu rests in a ripening patch of this grass.

Streamside Saw Grass

Arthraxon prionodes   Length 40–60cm

Description: A distinctive, pretty grass that grows
in a trailing, untidy fashion from a hidden base, often
rooting along the nodes as it scrambles over rocks and
streamside vegetation. Leaves, 2–7cm long, sheathe
the base of the stem. They have a striking elongate-oval
shape and are bright yellow-green with reddish-purple
margins and markings. The inflorescence consists of
several finger-like spikes clustered together at the end of
a long, thin stem, which is often slightly bent to one side
or arched. Spikelets are covered with short silvery hairs.

Distribution & Habitat: A common and widespread
grass. It typically grows among rocks along the edges
of streams and seasonal pools, and in damp spots in
woodland and glades. Found in areas 1,000–2,000m
above sea level.

Ecology & Uses: A localised grass that is not typically
grazed by livestock or wildlife. With its fairly tough
leaves and thin, wiry growth form, it does not have
a high nutritional value. This grass can also grow in
shallow soil and rocky crevices, making it an important
species for stabilising microhabitats along the edges
of streams. The attractive leaves and growth form offer
potential for use as an ornamental plant.

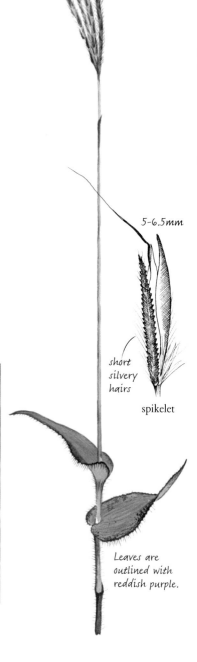

5-6.5mm

short
silvery
hairs

spikelet

Leaves are
outlined with
reddish purple.

Leaf bases appear to be
wrapped around the stems.

The young inflorescence
has a pinkish tinge.

Rocky streams are a typical habitat of this grass.

Sweet-pitted Grass

Bothriochloa insculpta   Height 25–200cm

Other name: Pinhole Grass

Description: An elegant perennial grass, sometimes reaching over 2m. This erect plant has a striking yellow stem that holds up the inflorescence, which sways gently in a breeze. Leaves are 5–30cm long. Each inflorescence has a few ruddy brown racemes that fade with age. The spikelets have a distinctive 'pit', hence the common name.

Distribution & Habitat: A widely distributed grass throughout the region. It grows well on a wide range of soils and in varied rainfall conditions. Found in grassland and old cultivated areas, also in fallow and rocky areas from sea level up to 2,100m. Very common in many of the region's national parks, including the Masai Mara and Serengeti; in Laikipia county; and areas around the Great Rift Valley.

Ecology & Uses: An important grass when young for livestock and wildlife. As the plant ages, the leaves accumulate an essential oil that limits grazing by some species but is attractive to others, including white rhinos, whose trampling encourages dense, mat-like growth. As a fast-growing species that can form extensive stands, it plays an important role in stabilising soils in overgrazed areas.

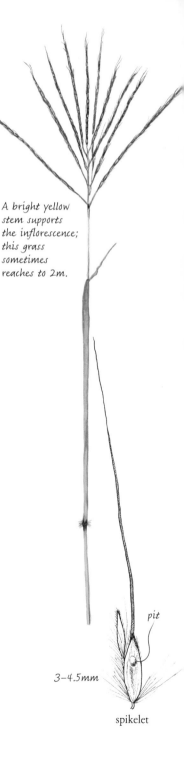

A bright yellow stem supports the inflorescence; this grass sometimes reaches to 2m.

pit

3–4.5mm

spikelet

The flowering panicle is held up by a bright stem.

Ripening spikelets turn from purple to brown.

A stand of Sweet-pitted Grass coming into seed

Tanner Grass

Brachiaria arrecta   Height 50–135cm

Description: A sprawling, untidy perennial grass that grows in a prostrate manner, rooting at the nodes. It can form dense mats, especially where it is managed as a pasture or fodder species. Leaves, 3–25cm long and up to 15mm wide, are typically quite soft and juicy. The inflorescence is a straight panicle with short branches. The spikelets are round, smooth and borne in two rows.

Distribution & Habitat: A very widely distributed grass across the region. More typically found in higher-rainfall areas where it grows in woodland and fallow land, along paths and forest edges, and on roadside verges. In the wild it often grows around the edges of wetlands or swamps, and may also be found in stagnant water. Mostly found in areas 400–2,000m above sea level.

Ecology & Uses: Developed and managed for its nutritious, leafy yield, this grass is an important fodder species for livestock, and a good grass for dairy cattle (when managed properly). In the wild it is also grazed by livestock and the semi-aquatic sitatunga antelope. It can tolerate both saline and acidic soils in wetter areas.

A dik-dik grazes on young growth of this grass.

Reddish spikelets appear some time after the leaves.

3–4.5mm

spikelet

Tanner Grass in the cleft of a tree

In flower, two rows of spikelets are visible.

A dense carpet of this grass is typical of growth in the shade of trees.

Palisade Signal Grass

Brachiaria brizantha   Height 30–200cm

Description: An elegant, loosely tufted perennial grass that grows both in individual clumps and as a dense, leafy sward covering the ground. Leaves are a vivid light green, 10–100cm long. The inflorescence is a distinctive arrangement of 1–16 spike-like racemes held almost at right angles from the main stem. The plump, round spikelets are neatly arranged in rows, bright green in colour and often with purple markings.

Distribution & Habitat: A common and widespread grass throughout East Africa. It is found in wetter sites and on better soils in woodland, bushland and grassland, and in fallow and recently cultivated land; also found on roadside verges and in pastures. Grows in areas 300–2,500m above sea level.

Ecology & Uses: An important and nutritious drought-tolerant grass for grazing by livestock and wildlife. Herbivores in natural pastures favour this grass and rapidly consume it when it is available. The species has been widely introduced in the tropics for use as pasture, hay and silage. In Brazil, it has been developed, through plant breeding and selection, into several useful varieties; subsequently it has been grown in other countries. Considered good fodder for rabbits as well as in zero-grazing, livestock-rearing systems.

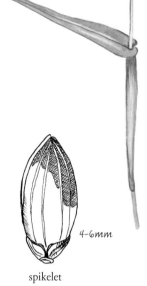

Racemes point away from the stem at an angle and can vary in number.

Ripening inflorescence

4–6mm

spikelet

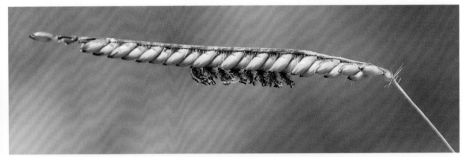
Close-up view of neatly arranged spikelets on a raceme

A spreading clump of Palisade Signal Grass offers good grazing.

Sweet Signal Grass

Brachiaria eruciformis Height 10–60cm

Description: A short, modest annual grass that grows in loose, often single, tufts in a slightly flat, spreading manner, occasionally reaching a height of 60cm. Leaves, 2–17cm long, are soft and bright green, often tinged with reddish-purple markings. The inflorescence is held upright, often at a slight angle, projecting outwards from the base of the plant. It is a delicate arrangement of narrow, finger-like stems with spikelets neatly packed together in tight rows.

Distribution & Habitat: A common and widespread grass in the region. It thrives in slightly damp or wetter areas, including on black-cotton soils and clays. Found at altitudes from 500–2,500m above sea level. Also widespread in other parts of tropical and southern Africa.

Ecology & Uses: An important source of grazing for livestock and wildlife, and used as a fodder grass in some places. In the dry season it is sought out for its nutritious leaves. It is a pioneer species and can also grow in wetter areas that retain moisture well into the dry season. As a fast-growing annual, it provides ground cover and stabilises areas of bare ground, thereby reducing soil erosion.

The fine, delicate inflorescence grows on a straight stem.

1.7–2.7mm

spikelet

This grass helps stabilise bare ground.

A ripening inflorescence growing at an angle

Young spikelets are tightly packed together.

Hairy Signal Grass

Brachiaria lachnantha Height 20–80cm

Description: A densely tufted, elegant perennial grass that forms dense stands. The basal sheaths encircling and supporting the stems and leaves are covered in silky hairs. Leaves are flat, 5–40cm long. The inflorescence has a tall, erect spike held at a slight angle above the plant. Spikes are light green with silvery hairs that fade as the plant ages, making them soft and silky to the touch. Spikelets are arranged along a series of slightly angled spikes that are all typically pointed in one direction on the inflorescence.

Distribution & Habitat: A common and widespread grass in East Africa and into the Greater Horn of Africa region. It thrives in seasonally waterlogged conditions. It is often found in association with Red Lintonia and wire grasses (*Pennisetum*). It is one of the dominant grasses of the black-cotton clay soil habitat. Widespread from sea level to 2,000m in altitude.

Ecology & Uses: An important, nutritious grass that is grazed by livestock and wildlife. Cattle and zebra feed on this grass in the dry season as they range in the black-cotton clay soil areas. Also utilised by eland, elephant and other herbivores. Seeds are eaten by birds.

Spikes are light green with fine silvery hairs that shimmer in the sunlight.

Ripening infloresences turn straw coloured with age.

3-4.5mm

spikelet

A stand of Hairy Signal Grass coming into flower; the hairy inflorescences are readily spotted from a distance.

Hairy Signal Grass ■ 49

Dryland Signal Grass

Brachiaria leersioides   Height 10–100cm

Other name: Blue Signal Grass

Description: A short to medium-sized annual grass
that forms loosely spaced and sprawling growth,
sometimes trailing slightly over the ground, rocks and
other plants. The plant is rather hairy, with long bristly
hairs on the stems and leaf bases. Leaves are flat,
5–20cm long. The inflorescence consists of a number
of racemes spread out along the main stem and
typically pointed down at a slight angle when mature.
Racemes are a distinctive light green when young,
fading to a straw colour as the plant ages.

Distribution & Habitat: A widely distributed grass
in the drylands and semi-arid regions of East Africa. It
thrives on rocky ground and is an important component
of the drylands grass community. Occurs from sea level
up to 1,800m.

Ecology & Uses: An important grazing species for
both livestock and wildlife in arid and semi-arid regions.
The young leaves are especially sought after by sheep
and goats, and are also nibbled selectively by dik-dik
and klipspringer.

*Racemes point
away from the
main stem, and
are often angled
downwards.*

Young flowering inflorescence

2–3.5mm

spikelet

Dryland Signal Grass can form stands on its own or grow among other grass species in drylands.

This grass establishes on bare ground, helping to stabilise soils.

Forest Broom Grass

Brachypodium flexum   Height 30–100cm

Description: A slender, perennial grass that trails and droops over neighbouring plants; its leaves cluster together, forming a dense mat or clumped tussock. Stems are long and weak, sometimes reaching up to 100cm in height. Leaves, 5–20cm long, are flat, thin and slightly rough on the underside. The inflorescence is often slightly drooping, with an open arrangement of overlapping spikelets. Awns are short and wiry.

Distribution & Habitat: A fairly common and widespread grass in East Africa. Found exclusively in highland forest and moorland glades and along streams in high-altitude zones 2,000–3,000m above sea level. It also occurs in other parts of tropical Africa and in southern Africa.

Ecology & Uses: Not heavily grazed as it is fairly sparsely leafed. In highland forest glades and edges, it is an important food source for various species of antelope including duiker, bushbuck and buffalo. As part of the forest understorey, it can grow in shady spots and helps provide ground cover.

The sparse, delicate and drooping inflorescence is sometimes difficult to spot in the undergrowth.

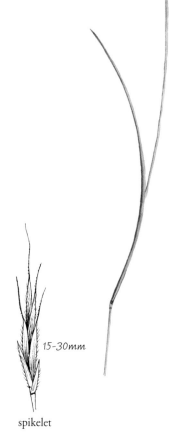

Flowering inflorescence

15-30mm

spikelet

Maturing inflorescence

Forest Broom Grass can form stands along shaded paths through highland forest.

Rescue Brome Grass

Bromus unioloides   Height 10–80cm

Description: An elegant, erect annual grass that grows in loose, graceful tufts. The leaf sheaths are soft and hairy. Leaves are bright green and soft to the touch, 5–30cm long. The inflorescence is a pretty, distinctive open panicle, 10–40cm long. The spikelets hang from the spike in a loose, drooping manner, each spikelet with a flattened, slightly tapered shape arranged in an overlapping pattern. Spikelets are a soft green with pinkish hints when young, turning paler as they ripen.

Distribution & Habitat: A common and widespread grass in the highland areas of East Africa. It thrives at the edges of fields and hedgerows, in fallow land, and on roadside verges in highlands, 2,300–3,000m above sea level. Originally introduced from South America, this grass is now found in both natural and cultivated areas throughout the world.

Ecology & Uses: Considered a good grazing grass for livestock in highland areas. Growing on roadside verges and communal lands, this grass provides food for tethered and herded animals in mixed-farming systems. Can be cultivated and managed for pasture in wetter, cooler highland zones.

Panicles are large with loosely arranged spikelets that rustle in the slightest breeze.

15-25mm

spikelet

Young spikelets appear slightly flattened.

Spikelets emerging from the leaves are pale green.

Mature inflorescence starting to ripen

This grass often grows at the edges of fields, near tree trunks and along hedgerows.

Bush Grass

Calamagrostis epigejos　　　Height 60–150cm

Other names: Common Wood Reed Grass,
Feathertop Reed Grass

Description: An elegant and robust perennial grass
that grows in tall, dense stands from a hidden, creeping
rhizome. The plant typically grows over a metre tall in
most areas. Leaves are flat and slightly rough, up to
45cm long. The inflorescence is an attractive feathery
panicle, ranging from green to pink when young, fading
to a pale creamy brown. The tiny spikelets, 6–8mm long,
are densely packed together. A beautiful, striking grass
when in flower and seed, making it easily recognisable
from a distance.

Distribution & Habitat: A common and widespread
grass in the highland and montane regions of East
Africa. It often grows in areas with moisture, including
along drainage lines and edges of wetlands, and in
seasonally flooded glades or grassland. It grows in
highland areas 2,000–3,000m above sea level. Also
widely distributed in all temperate regions of the world.

Ecology & Uses: Not heavily grazed by wildlife owing
to its relatively tough leaves. As this grass often grows
in areas that are part of the watershed, its dense stands
provide important habitat and nesting sites for birds. The
grass can be harvested for use as mulch or cultivated as
ground cover or an ornamental in gardens. This hardy
species can survive cold weather and being pruned; it
can also grow in saline conditions.

*Large feathery
inflorescences are
readily detected from
afar, appearing green
to pink when young.*

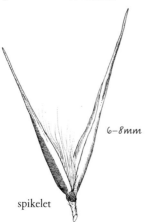

6–8mm

spikelet

A stand of Bush Grass in early stages of growth and flowering

The grass ripens to a coppery colour and forms dense stands in damp spots at higher altitudes.

Indian Sandbur Grass

Cenchrus biflorus   Height 5–90cm

Description: A stiff, upright annual grass. Leaves are flat and tapering, 2–25cm long. The inflorescence is a narrow spike, 2–15cm long. Spikelets are neatly packed together along the length of the spike. Individual spikelets have a slightly prickly surface with a short, sharp tip and are covered with barbs that readily attach to fur, feathers or clothing.

Distribution & Habitat: A common and widespread grass in the coastal, warmer and drier regions of East Africa. It grows on coastal sand dunes, roadside verges, recently disturbed ground and abandoned fallow land, from the shoreline to 140m above sea level. Also widely distributed in the Sahel and Greater Horn of Africa regions, and further afield in parts of Asia.

Ecology & Uses: Grazed when young, this is an important grass for both livestock and wildlife; the young leaves are nutritious, making it sought after by sheep, goats and camels. In parts of the Sahel, and in the drylands and deserts of India, this species is utilised as pasture and fodder for animals, and as food for people. It can be cut for hay and animal feed during its growth period, with the added advantage that it can be repeatedly harvested as it grows vigorously following rains. This pioneer species is important for stabilising sand dunes and disturbed ground.

Prickly spikelets are a uniformly bright green colour when young.

Prickly barbs cling to passers-by.

3.5-6mm

spikelet

Young and ripening
inflorescences

Maturing inflorescences with
darkening spikelets

Iñaki Abella Gutiérrez

The critically endangered hirola
grazes on this important grass.

This grass grows in dryland areas, alongside other grasses and thorny scrub.

Buffel Grass

Cenchrus ciliaris   Height 10–150cm

Other name: African Fox-tail Grass

Description: One of the most distinctive and recognisable of African grasses. This spreading perennial is extremely variable in height and grows in a trailing fashion, often forming large, dense mats. Flat leaves, 3–25cm long, grow in abundance. The pretty and spike-like inflorescence has a soft, fluffy appearance (hence the other common name, 'African Fox-tail Grass'). Spikelets are purplish green when young, growing darker with age and eventually deepening to a russet tone when the seeds are ripe.

The dense, feathery spikes change colour from green and purple-tinted when young to pale straw or grey as they age.

Distribution & Habitat: One of the most widespread and abundant grasses in East Africa and many other tropical regions in the world. It grows and thrives on a range of soils in woodland, bushland and open grassland, especially in rangeland and semi-arid areas at an altitude from sea level up to around 2,000m.

Ecology & Uses: One of the most important wild fodder grasses in Africa; it produces a good quality hay. The leaves are very nutritious and the growth pattern can support heavy grazing by many different animals. A favourite food for elephant, buffalo, eland, waterbuck, zebra and cattle. This grass can also be sown directly for restoration of degraded land. At the flowering stage, many different insects visit the inflorescence, which also hosts sap-feeding aphids. In wet years, the seeds may be infected by a smut fungus, which turns the seeds a sooty black colour and releases dust-like fungal spores when touched; it can kill the seeds if there is a heavy infestation.

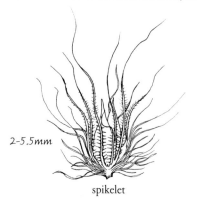

2-5.5mm

spikelet

A *Lipotriches* bee gathers pollen on this grass.

Inflorescence infested with a smut fungus

A stand of ripening Buffel Grass

Abyssinian Rhodes Grass

Chloris gayana   Height 50–200cm

Other names: Rhodes Grass, Boma Rhodes Grass

Description: A striking, erect perennial grass
that tends to grow in dense stands. It forms
stolons that grow densely, developing a thick, leafy
sward. Leaves are either flat or slightly rolled inwards,
15–50cm long. The distinctive inflorescence has a
fan-like shape comprising a variable number of spikes
clustered together and arched out from the base. This
digitate inflorescence grows on the end of the stems,
held above the leaves. The spikelets are brown and
shiny, 2.5–4mm long and with short awns.

*The inflorescence
grows on a
straight stem,
held in a fan
shape above
the leaves.*

Distribution & Habitat: A common and widespread
grass in East Africa. It grows wild in open grassland,
woodlands and pastures, and on roadside verges. It
can thrive at various altitudes from 120–2,500m above
sea level and on a range of different soil types. Widely
introduced and cultivated in many areas of the world.

Ecology & Uses: Considered one of the most useful
and valuable grasses around the world for livestock, it
is also cultivated for hay and in pastures. In East Africa,
it is widely grazed by both livestock and wildlife. As this
grass establishes rapidly, its seeds are harvested and
sold for direct sowing to create new fields; it can also be
useful in land reclamation, as a cover crop and to help
restore rangelands and soils. Abyssinian Rhodes Grass
has been developed into a number of useful varieties.

2.5–4mm

spikelet

Many insects, including ladybird beetles, thrive on this grass.

A sward of Abyssinian Rhodes Grass

A young inflorescence is easily recognised from a distance.

Ripe inflorescences are ready for harvesting and baling as hay for livestock.

Spiderweb Chloris

Chloris pycnothrix Height 15–55cm

Description: A short, sprawling grass that grows from a creeping rhizome and often forms dense mats. Leaves, 2–10cm long, are narrow and flat with a blunt tip. The spikelets have dense, closely pressed awns and are held on a star-shaped inflorescence, typically with 4 or 5 finger-like extensions. Spikelets are green with reddish awns when young, curving and turning straw coloured, and with the awns more spread out, as the grass matures.

Distribution & Habitat: A very widely distributed grass in East Africa. It grows in more open, short grassland, recently disturbed areas, on road verges and where animal trampling and grazing is more intense. Can be both an annual in more arid regions, and a perennial in wetter areas. Grows on a range of soil types and at different altitudes.

Ecology & Uses: An important grass for both livestock and wildlife. Several grazers that prefer short grasses, including white rhinos and warthogs, feed preferentially on this species. This tough grass can survive a degree of trampling and heavy grazing. It has a creeping growth pattern, making it a good species for stabilising bare soil and preventing erosion. It can be readily established by rooting the creeping runners or by sowing seed.

This important pioneer species can stabilise bare soil.

The reddish-tinged awns of the young inflorescences turn straw coloured and beige to silvery grey as they ripen.

2.5–4.5mm

spikelet

Young, freshly sprouted inflorescences

The inflorescences grow paler as they ripen, creating a dense and untidy mat that almost looks woven.

Feather-top Chloris

Chloris virgata Height 50–60cm

Description: An elegant, showy annual grass of medium height. It grows in tufts from a creeping base and forms dense stands. The leaves are flat and narrow, 10–30cm long. Leaf bases form sheaths with a distinctive keeled edge around the grass stems. The dense, feathery inflorescence consists of 5–15 spikes held open in a slightly fanned manner in the young fresh grass, becoming denser and more compact in older plants. Young spikelets are bright golden green in colour, with a distinguishing characteristic of darkening, then turning blackish as the grass matures.

Distribution & Habitat: A very common and widespread species in East Africa. It grows in a range of habitats, including wooded and open grassland and bushland, as well as in fallow, old cultivated areas or along roadsides. Feather-top Chloris grows on a range of soil types and at different altitudes.

Ecology & Uses: An important and useful grass for both livestock and wildlife. The dense tufts sprout back after grazing and the heavy seed production enables the species to readily establish itself on bare ground; these two factors make it an important grass for soil stabilisation and erosion control. The rapid growth following rains makes it an especially important source of fodder in arid and semi-arid regions.

Young inflorescence with spikes tightly clustered

Greenish-coppery spikelets change to black as the grass ripens.

2.5–4.5mm

spikelet

As the grass ripens, awns become more visible and spikes fan out.

Mature inflorescences turn black and start shedding seeds heavily.

A stand with both young and ripening inflorescences

Sickle Grass

Ctenium concinnum   Height 50–100cm

Description: An elegant perennial grass that grows in fairly dense clumps. Leaves are hard and narrow, with inward-folded margins, 10–30cm long. The distinctive inflorescence is a narrow spike, typically slightly curved, with spikelets neatly arranged in two compact rows. As the spikelets ripen and the seeds mature, the inflorescence becomes more tightly curled, almost circular, hence the common name 'Sickle Grass'. The spikelets are a pale green when young, with fine purple awns.

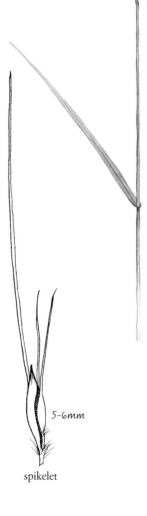

Bright pinkish awns on a curved inflorescence are distinctive.

Distribution & Habitat: A common and widespread grass in the lower-altitude bushland, forest and near-coastal regions of East Africa. It grows and thrives in open areas, glades and bushland, often appearing before the trees leaf out following the dry season or an extended period of drought. The species prefers sandy soils and red soils, in areas at an altitude of 100–350m.

Ecology & Uses: A hardy, drought-tolerant grass that is grazed by livestock and wildlife wherever it is available. It is important grazing for several dryland herbivores, including the hirola, a rare and endangered antelope in Kenya, and the dibatag, a slender antelope in Somalia. The seeds attract seed-eating birds and often host aphids tended by ants. The species provides ground cover and protects soils from erosion in areas with sparse rainfall.

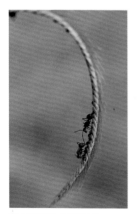

Polyrhachis ants tend aphids on Sickle Grass.

Young inflorescences form a curved shape.

5–6mm

spikelet

Stands of this grass have a slightly untidy appearance.

Narrow-leaved Turpentine Grass

Cymbopogon pospischilii   Height 40–100cm

Description: An elegant, often sparsely tufted perennial grass. Leaves are flat, 15–30cm long, and held at close angles to the main stem. The leaves and stems of young plants are often tinged with bright yellow. The inflorescence is a loose, open arrangement that is spread along the upper length of the plant. There are many paired or clustered short spikes arranged in a somewhat haphazard fashion. These are yellowish with reddish hints when young, becoming silvery and fluffier in appearance as they age and ripen.

Distribution & Habitat: A common and widespread grass of the mid-altitude and drier highland areas of East Africa. Typically found on slightly rocky soils or slopes in bushland, open grassland and in semi-arid areas, 1,000–2,500m above sea level.

Ecology & Uses: Like other turpentine grasses, the oil-rich leaves and stems of this species exude a turpentine odour, which helps to deter grazers. Some species of *Cymbopogon* are commercially exploited; the insect-repellent citronella oil is derived from Asian species of *Cymbopogon* such as Lemon Grass (*C. citratus*). There is anecdotal evidence that herding livestock through stands of these grasses helps to control both ticks and flies; as the animals brush against the grasses, the oil is transferred to their skin, where it acts as a repellant to pests and parasites.

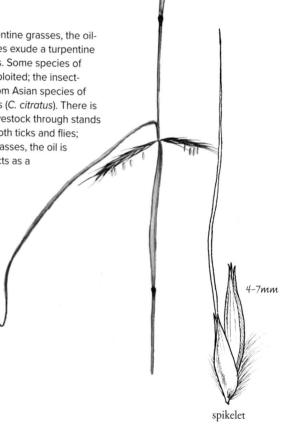

Reddish spikes are arranged in loose clusters up the stem.

4–7mm

spikelet

This grass occurs in higher-altitude regions in East Africa.

Young inflorescences are tinged purple.

A bushbuck moves through a stand of ripening grass.

This grass typically grows in disturbed areas alongside species such as Buffel Grass and Shining Top Grass.

Couch Grass

Cynodon dactylon

Height 5–40cm

2–2.5mm

spikelet

Description: A short, mat-forming grass that grows in dense stands from a creeping rhizome base; leaves are borne in tufts and from branching stems along creeping stolons. Flat or folded leaves are soft and juicy, 2–10cm long. The inflorescence typically consists of 4–6 spikes spread out in a star-shaped arrangement. It has a purplish colour when young, fading as the spikelets mature and dry.

Distribution & Habitat: This very widely distributed grass forms dense, monodominant stands and mats that are visible as natural 'lawns' from a distance. It grows on a range of soil types at different altitudes, from sea level to areas over 2,000m high.

Ecology & Uses: Grows rapidly following rains. An important grazing species utilised by both livestock and wildlife in East Africa. The dense mats are favoured by grazers that like short grasses. It is also widely used as a lawn grass. Seeds are eaten by many different birds. Bees, including honeybees, gather the pollen from this species during the flowering stage. The robust growth from both rhizomes and stolons makes this species a great soil stabiliser, preventing soil erosion.

The star-shaped inflorescence consists of 4–6 spikes.

Couch Grass in flower

Heavily grazed areas become natural lawns and are favoured by livestock and wildlife.

Boran cattle graze on Couch Grass.

Star Grass

Cynodon nlemfuensis   Height 20–60cm

This grass has a spreading and whorled inflorescence.

Description: A robust, short to medium-sized perennial grass that grows in a dense tufted pattern from creeping runners that spread out along the surface of the soil. The leaves are flat, soft and juicy, 5–16cm long. The inflorescence is star-shaped and consists of a variable number of spikes arranged in a whorl. These whorls are slightly curved and a dark reddish colour when young.

Distribution & Habitat: A common and widespread grass in East Africa. This species grows in a wide range of natural habitats, including wooded grassland, savanna, open grassland and bushland, as well as in recently disturbed areas and on fallow land. Forms dense leafy stands, especially where good soil and moisture conditions allow. Found from sea level up to areas 2,500m in elevation.

Ecology & Uses: An important grazing species for both livestock and wildlife in East Africa. Its robust, leafy growth and ability to sprout new growth from the runners make this grass very useful in both managed pastures and natural systems. The habit of growing outward from stolons is very important for stabilising soil and preventing erosion. This species can be cultivated from runners as well as from seed.

Star Grass in flower

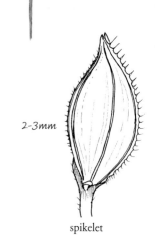

2–3mm

spikelet

Ankole cattle resting on a field of Star Grass

This grass is a favourite of Grevy's zebra and impala.

Mixed feeders, here Grant's gazelle, make use of this grass.

Elephant spend a great deal of time pulling up runners of this grass using their trunks.

Common Crowfoot Grass

Dactyloctenium aegyptium   Height 20–40cm

Description: A distinctive, attractive, short annual grass that can reach up to 70cm in height. It grows in dense mats from a creeping stolon, often in a robust pattern. Leaves are flat, 3–25cm long. The inflorescence is a striking, star-shaped 'crow's foot' that consists of a variable number of spreading spikes held outward from the central stem, typically borne on stalks above the plant. Spikelets are bright green when young, fading to light brown.

Distribution & Habitat: A very common and widely distributed grass in East Africa. Typically found in woodland, bushland and open grassland habitats, and on road verges and recently disturbed ground. It grows on a range of soil types, from sea level up to areas over 2,000m above sea level.

Ecology & Uses: An important fodder grass for both livestock and wildlife, especially in arid and semi-arid regions. The leaves are highly nutritious for grazers. Sheep are particularly fond of this plant and will graze on it voraciously. Its seeds are used by many different birds and also by harvester ants (*Messor* spp.).

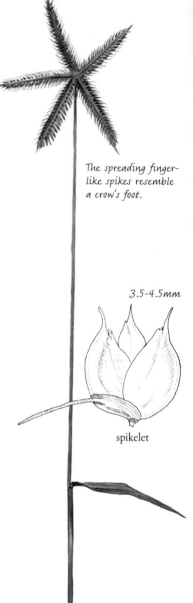

The spreading finger-like spikes resemble a crow's foot.

3.5–4.5mm

spikelet

The striking, star-shaped inflorescence makes this one of the most distinctive grasses.

This grass can form dense stands, with a mix of flowering and ripening inflorescences.

Small Crowfoot Grass

Dactyloctenium geminatum   Height 30–100cm

Description: An attractive, robust and tough grass that grows in dense mats from strong, wiry stolons and can reach up to 100cm in height. Leaves are narrow, flat and compact, 4–25cm long; they can be smooth or slightly hairy to the touch. The distinctive inflorescence consists of single or paired short, stiff spikes that are held out at an angle from the stems. Spikes are 3–7cm long, bright green when young, then fading and breaking up as they ripen and age.

Distribution & Habitat: A common and widespread grass in the coastal regions of East Africa. It grows on dunes along the seashore, in glades and moist settings in coastal bush, in fallow land, and within successional vegetation communities in coastal forest and woodland. It thrives from sea level up to areas 200m in elevation.

Ecology & Uses: Grazed when young by livestock and wildlife. Many different birds and insects live and nest within the stands of this species. As a robust, mat-forming and tough grass, it is extremely important for providing ground cover, stabilising sand dunes and preventing soil erosion.

The spikes are held out at a wide angle, and are bright green when young.

Young inflorescence before flowering

3-5.5mm

spikelet

Inflorescences turn the colour of straw as they ripen.

A stand of Small Crowfoot Grass

Small Crowfoot Grass ■ 79

African Couch Grass

Digitaria abyssinica
(= Digitaria scalarum)

Height 5–100cm

Description: A tough, wiry, creeping perennial grass that branches out from rhizomes and can form dense mats or clumps. Leaves are 3–15cm long, flat and variable in length, often forming a dense carpet in areas where the plants are heavily grazed. The inflorescence is an erect, compact panicle, dark purplish or green when young, with individual spikelets clustered on a series of spreading spikes. Often many individual inflorescences are borne in parallel along the length of the mats.

Distribution & Habitat: A very common and widespread grass throughout East Africa. As a weedy yet tough species, it is found in a wide range of habitats, but is more common in recently disturbed ground. It grows from sea level to around 3,000m. The species is also found in other parts of Africa and the wider tropics.

Ecology & Uses: An important grazing species for both livestock and wildlife. As a tough pioneer species, this grass is utilised by livestock in mixed-farming systems, especially where livestock are rotated through fallow land or tethered to graze on roadside verges. The species is very important at higher altitudes, where it provides ground cover and helps to prevent erosion.

Inflorescence tends to be darker at higher altitudes.

2–2.5mm

spikelet

At high altitudes, this grass has much shorter inflorescences, flowering close to the ground.

Flat leaves have a soft, slightly hairy feel.

African Couch Grass often grows in association with clovers; it forms dense, palatable mats for grazing.

Silver-haired Finger Grass

Digitaria argyrotricha   Height 30–50cm

The fine, delicate inflorescence consists of narrow 'fingers'.

Description: A creeping annual grass that typically grows in a dense manner, forming a short to medium-sized mat-like stand. Variable in height depending on the location, it may reach 170cm in more shaded or wetter sites. Leaves are 2–10cm long, but typically 5–6cm, flat and narrow with a fine, pointed tip. The inflorescence is fine; it typically consists of 2–5 narrow 'fingers' held above the plant on a narrow stem. Spikelets are 1.5–2.4mm long, with a few pale silvery white hairs visible when viewed closely. The specific name is derived from the Greek, meaning 'silver haired'.

Distribution & Habitat: A common and widespread grass in the coastal, bushland, hinterland and lower-altitude regions of East Africa. It grows and thrives on both sandy and red lateritic soils. Found from sea level to around 500m.

Ecology & Uses: A locally important source of grazing for both livestock and wildlife. Significant for seasonal grazing in areas where grasses are often sparse, especially in bushland habitats following rains. It often grows in stands mixed with other grasses, including Red-seed Guinea Grass (*Panicum atrosanguineum*). The seeds are eaten by birds such as crested and vulturine guineafowl. It is an important ground cover on fallow land within shifting cultivation zones in bushland areas.

1.5–2.5mm

spikelet

A dense mat of leaves and stems is typical of this grass following good rains.

The creeping runners help to cover and stabilise bare, sandy soils.

Dense mats of this grass grow in stands mixed with young Tall Guinea Grass in coastal forest.

Hairy Finger Grass

Digitaria ciliaris   Height 20–100cm

Description: An annual grass of variable height with a straggly growth pattern. Leaves are 3–25cm long, flat, borne closer to the lower half of the plant. The inflorescence consists of several spike-like racemes arranged in a graceful cluster, pointing upwards. Spikelets are closely pressed together and typically covered with silvery white hairs. This is a highly variable species with a number of subspecies that have been described.

Distribution & Habitat: A widely distributed and common species across East Africa and many of the tropical regions of the world. This is a weedy pioneer grass that grows on disturbed ground, recent fallow, rocky areas and roadside verges. It often forms sparse clumps and grows interspersed with other grasses. Found from sea level up to 2,000m.

Ecology & Uses: As part of the mixture of dryland grasses, this species is moderately grazed by both livestock and wildlife when young. Its ability to establish on disturbed ground and to grow rapidly from seed makes it an important species for stabilising soils and preventing erosion in rangelands.

The tall, spindly racemes have a slightly untidy appearance and readily break off when ripe.

2.4–3.8mm

spikelet

This grass is an important pioneer species on bare ground.

Young Hairy Finger Grass in early stages of flowering; hairs are visible on the stem below the inflorescence.

Ripening Hairy Finger Grass grows along a roadside verge.

Feeble Finger Grass

Digitaria debilis   Height 20–60cm

Description: A straggly, thin, frail-looking annual grass. The plant often roots along the stems and grows in sparse clumps, with a few stems rising up from the ground. Leaves are tapered, 3–15cm long. Hairs grow along the margins and stem sheaths. The inflorescence is made up of a variable number of spikes, held in a slightly open triangle, ranging in length from 3–16cm. Individual spikelets are spread out along the stems, held close and with a slightly spiky appearance, readily detaching when the plant is handled.

Distribution & Habitat: A fairly widespread grass in East Africa, but typically localised in its occurrence. It grows and thrives on poor, sandy soils in seasonally wet areas, in pans overlying rocks, and at the edges of rocky outcrops. This grass grows from sea level up to 2,000m. It is also found in other parts of Africa, including Madagascar, and the Mediterranean.

Ecology & Uses: Typically this grass is not heavily grazed by livestock or wildlife, though it is grazed when other grasses have dried up or been grazed down to the ground. It grows in conditions where relatively few other plants thrive, making it an important pioneer species that provides stability and ground cover.

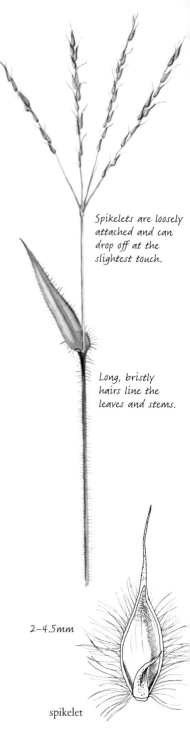

Spikelets are loosely attached and can drop off at the slightest touch.

Long, bristly hairs line the leaves and stems.

2–4.5mm

spikelet

Young Feeble Finger Grass and other species sprouting among rocks early in the rainy season

This grass forms small stands in sandy soils; flowers appear late in the rainy season.

Soft hairs on the leaves and stem glisten in sunlight and trap dew.

Feeble Finger Grass ■ 87

Shade Finger Grass

Digitaria gymnotheca Height 30–60cm

Description: A trailing, straggly perennial grass that often grows in a creeping fashion, with stems rising up to 60cm. Bright green leaves are 5–20cm long, flat and soft, with a tapered pointed tip. The inflorescence has a variable number of spikes, held in a slightly open triangle and ranging in length from 7–15cm. Individual spikelets are spread out along the stems, held closely together in pairs.

Distribution & Habitat: A fairly widespread grass in East Africa, but typically localised in its occurrence. It grows and thrives in shade in coastal and dryland regions. The species can form small, dense carpets in areas where there is sufficient moisture. It does well on both sandy and rocky soils, from sea level to 400m.

Ecology & Uses: Important grazing locally for both livestock and wildlife. It is often in leaf well into the dry season, offering fresh grazing when other grasses have dried. Many insect species take refuge in this grass; ants may be found on it, tending aphids or other hemipteran bugs for honeydew. This grass also grows in gutters, ditches and around houses, as long as moisture is available.

2.5–3mm

spikelet

Long, narrow leaves taper to a fine point.

The loose, open inflorescence can be difficult to see in dappled shade.

Stems of this grass often hold clusters of aphids, which are tended by ants.

Yellow anthers are clearly visible at the flowering stage, making the grass easier to identify.

This grass is an important microhabitat for many insects and other creatures because it grows in the shade and stays green for long periods.

Milanje Finger Grass

Digitaria milanjiana   Height 50–250cm

Other names: Woolly Finger Grass, Digit Grass

Description: A striking, robust grass that grows in dense, beautiful stands. It tends to grow taller in wetter habitats, including in moister zones closer to the coast, with stems rising to an impressive height. Leaves are 15–30cm long, narrow and flat. The very variable inflorescence is made up of 4–18 stiff spikes, 5–25cm long. The fanned arrangement of the spikes is striking when this grass is in flower or seed.

Distribution & Habitat: An extremely common and widespread grass across East Africa. The species thrives in open grassland, bushland, woodland and forest glades, and can do well on many different soil types, including seasonally wet black-cotton clay soils. The species grows from sea level to over 2,000m. It is also found in many other regions of tropical Africa.

Ecology & Uses: An important source of nutritious grazing for both livestock and wildlife. The dense leafy stands are able to retain some moisture and nutrition into the dry season, making this grass a favourite for buffalo, elephant and other species that will often feed on it in succession, with the larger herbivores feeding on it first, then the smaller ones once new growth has sprouted. It makes excellent hay and is managed for fodder in some areas.

The large, wavy inflorescence is held high on a tall stem, with the fanned spikes visible.

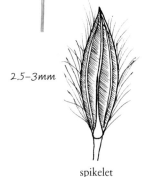

2.5–3mm

spikelet

This grass turns a light russet-coppery colour as it ripens, making it one of East Africa's prettiest grasses.

A dense stand of Milanje Finger Grass in a coastal forest glade provides grazing for many herbivores.

Long-plumed Finger Grass

Digitaria velutina   Height 15–80cm

Description: A short, sparse annual grass that grows in a slightly spread-out fashion. The stems are often bent over and will readily root at the nodes. Leaves are 2–15cm long, soft and fairly broad, with a tapered tip. The inflorescence consists of a number of spike-like racemes arranged around a short central stem.

Distribution & Habitat: A common and widespread grass species throughout East Africa and most of the continent. It typically grows in grassland, bushland and recently cultivated or fallow land, on a range of soil types and at different altitudes, from sea level to 2,500m.

Ecology & Uses: The nutritious leaves are widely grazed by both wildlife and livestock, especially in dryland areas. Selective herbivore species such as dik-dik, bushbuck, kudu and eland will seek out this grass and quickly nibble off the leaves, causing the seeds to drop to the ground, where they last in the soil for many years while waiting for the rain needed to germinate and establish new plants. The species grows rapidly once rains have fallen.

The inflorescence rises above the leaves, but the grass is easy to miss when not in flower.

1.5–2.5mm

spikelet

Young inflorescences are bright green with a variable number of racemes.

Readily establishing in fallow areas, Long-plumed Finger Grass is an important pioneer species.

Catstail Vlei Grass

Dinebra retroflexa Height 20–30cm

Description: A short, low-growing, sparse annual grass that often grows in a trailing fashion or forms small isolated clumps. Individual runners can spread along the ground for over a metre. Stems are robust, often dark purple. Leaves are flat, 5–16cm long. The inflorescence is a short, graceful spike with individual green and reddish-purple spikelets arranged along its length, loosely spread out, giving the impression of a fluffy cat's tail.

Distribution & Habitat: A widespread grass in East Africa and across the tropical regions of Africa, including the Greater Horn. It typically grows in open grassland and wooded grassland habitats, including on black-cotton clay soils. The species can grow and thrive in flooded ditches, on roadside verges and along the edges of seasonal pools or vleis. It is found from 300–2,000m.

Ecology & Uses: A small and compact species with moderate importance for grazing, being seasonally grazed as animals move closer to wetter areas. It is an important part of the ecosystem because it grows in a moist microhabitat where not many other grasses can survive, because of the waterlogged soils. This species serves as a perch for dragonflies and damselflies, as well as a shelter for young amphibians and aquatic insects.

Young, freshly sprouted inflorescences are dark, with hints of rich green.

The colourful inflorescences are easily spotted from a distance as they ripen.

5.5–9mm

spikelet

This grass is a common feature of flooded ditches and pans in black-cotton clay soils throughout the region.

Short Water Grass

Echinochloa brevipedicellata   Height 30–200cm

Other name: Swamp Rice Grass

Description: An elegant, robust grass that grows from trailing stems. It is often floating or partly submerged in water, with the leaves and flowering parts held erect above the stems. Leaves are 10–30cm long and soft, bright green with a rough feel to the touch. The inflorescence is an attractive, open panicle with the spikelets arranged on short stalks. Inflorescences are a reddish colour when young, deepening as they ripen. Individual spikelets are round and arranged in tight rows.

Distribution & Habitat: A widespread and locally abundant grass throughout the East African region and in many other tropical regions of the world. As its common names suggest, this grass grows along streams and riverbanks, and in seasonally flooded pans and wetlands. It is found from 400–2,000m.

Ecology & Uses: A nutritious and juicy grass that is eaten by both livestock and wildlife. It is an important seasonal forage and source of grazing for many different species, especially when other grasses have been grazed down or dried up completely. The rhizomes are eaten by baboons, rodents and warthogs as the water level drops in the wetlands and seasonal pans become dry. This species plays an important part in the stability of wetlands and swamps. It also has the potential to be developed as fodder for livestock.

The bright red coloration of the inflorescence makes this a striking grass when flowering.

Leaves feel rough and scaly.

1.5–2.5mm

spikelet

Young inflorescences have bright reddish spikes.

Bare, muddy ground is one of the places where Short Water Grass establishes and provides important stability.

Loose clumps grow alongside flooded areas, water pans and areas with slow-moving water.

Jungle Rice Grass

Echinochloa colona   Height 10–100cm

Description: A robust yet elegant, erect annual grass. Leaves are 5–35cm long, flat and slightly soft to the touch, sometimes with purple markings, making this grass similar to the related Antelope Grass (*Echinochloa pyramidalis*). The inflorescence is held above the leaves – a panicle of finger-like spikes, bright lemony green when young, decreasing in size as they ascend the main stem. The spikelets are densely packed together in rows, with each spikelet 2–3mm long.

Distribution & Habitat: A common and widespread grass in East Africa, and in many other regions of the tropics. It grows in damp areas in grasslands, along streams and riverbanks, at the edges of roads, and around seasonal pans and wetlands. Found from sea level up to 2,000m.

Ecology & Uses: Important grazing for both livestock and wildlife. This grass has high leaf production, which makes it useful for fodder. It also grows as a weed in flooded rice fields, where it competes with the cultivated crop. In some areas, particularly near coastal lowlands, it is an important food source for hippo, buffalo and other herbivores. As a fast-growing species that can establish while wetlands are drying up, it does provide nourishment when other grasses have already been grazed or become too dry for grazing.

The dense spikes and seeds can be harvested and eaten as is, or milled. This grass is the wild ancestor of the cultivated Sawa Millet.

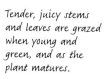

Tender, juicy stems and leaves are grazed when young and green, and as the plant matures.

Young, bright inflorescences grow gradually, rising above the leaves.

1.5–3mm

spikelet

Many different creatures, such as this freshwater snail (*Bulinus* sp.), find refuge and sustenance among the stems and roots of this grass.

A common feature of lakes, ponds, dams and wetlands, this weedy grass is a fast-growing species.

Antelope Grass

Echinochloa pyramidalis   Height 1–4m

Description: A striking, robust, perennial reed-like grass that grows from a rhizome, with erect stems reaching up to 4m. Leaves are 8–60cm long, dense and slightly stiff, often marked with purplish stripes. The inflorescence is variable, ranging in length from 8–40cm, but typically 15–25cm long. A series of short spikes is angled away from the main stem, typically pointing slightly upwards. Spikes are green when young, but turn bright red as they mature, especially when the plants grow in full sun. The spikelets are densely packed together in tight rows.

Distribution & Habitat: A common and widespread grass in East Africa. It is found in swamps, wetlands, damp grasslands and seasonally flooded areas, and along slow-flowing seasonal streams, from sea level to areas above 2,000m. The species is also widely distributed across Africa and has been introduced to many other parts of the world.

Ecology & Uses: An important grass that is used by both livestock and wildlife. It is grazed when young, and while other grasses are drying with age. It is one of the grasses that are grazed by hippos as it grows in and around water. Because this grass is easy to establish along the edges of water, and can survive both flooding and drought, it is very useful for providing stability to newly built compacted earth dams and water pans. It also helps to aerate water and provides habitat for numerous insects, birds and amphibians. In addition, this species can be managed and cut for fodder.

Reddish inflorescences with densely packed spikelets stand tall above the plant.

Leaves have neatly spaced purple markings and feel stiff.

2.5–4mm

spikelet

Purplish stripes on the leaves make it easier to identify.

This grass grows in dense stands and is often in leaf year-round, providing important grazing.

The inflorescences change from green to bright red as they mature.

Antelope Grass can grow in seasonal rivers, providing an important refuge for many creatures and shelter for young fish and insects.

Panic Veldtgrass

Ehrharta erecta   Height 30–100cm

Description: A trailing, loosely tufted annual or perennial grass that often grows in a creeping fashion, draping itself over other grasses and vegetation. Highly variable, it tends to grow in a more straggly manner in shady locations. Leaves are 6–18cm long, flat, soft and pliable, and bright green. The inflorescence is a skinny, slightly untidy erect panicle, almost spike-like, 3–20cm long. Individual spikelets are packed very closely along the length of the panicle; pale green and oval in shape, they are 5–6.5mm long. Bracts may persist after the seeds have been shed.

Distribution & Habitat: A very common and widespread grass in the wetter highlands and upland regions of East Africa. It typically grows at the edges of hedgerows, in fallow areas on tree lots, and in agroforestry plantations, forest glades and woodlands. The species thrives at 1,500–2,700m above sea level.

Ecology & Uses: Important grazing for both livestock and wildlife. The soft, juicy leaves are often sought out by herbivores such as hares and by livestock that have been tethered to graze or confined in paddocks; the seeds are eaten by birds. Because this grass is easily established in shaded, moist areas, it is a popular ground cover for protecting soils from erosion due to heavy rain or water runoff.

The narrow, spindly inflorescence often sheds its spikelets, leaving empty bracts lining the stem.

Soft, juicy leaves are grazed by many creatures.

5–6.5mm

spikelet

Pale yellow anthers dangle from the spikelets.

Panic Veldtgrass can grow and thrive in shaded areas; leaves emerge through leaf litter in small tufts.

Finger Millet

Eleusine coracana　　　　Height 10–150cm

Other name: Goose Grass

Description: An attractive and striking annual grass that grows in dense, compact clumps with the leaves and stems clustered together, often fanning out from the base. Leaves and stems are hard and wiry. The plant is very difficult to uproot as it has a robust, deep root system. The folded leaves are 9–50cm long. Growing in the wild, this plant tends to be shorter and more compact, whereas cultivated forms are taller with a looser growth shape. The inflorescence consists of several finger-like spikes varying in number from 1–11. Spikelets on the wild form are arranged in tight rows; pale green when young and turning straw coloured as they ripen.

Distribution & Habitat: A very common and widespread grass in East Africa. It is often found growing wild along roads and paths, also in the ridge between the wheel ruts on dirt roads or bush tracks, and in fallow or recently cultivated areas. It does well on a wide range of soils and in seasonally arid zones. Finger Millet grows and thrives from sea level up to altitudes of 2,000m. The species has also been introduced in many other parts of the world.

Ecology & Uses: Because it is tough and easily established, the wild form is an important pioneer species, helping to stabilise soil and provide ground cover. It can grow on sand and rock, and helps other plants to establish while providing shelter for their seeds. Finger Millet seeds are eaten by a wide range of birds and rodents. Cultivated forms have larger seedheads. Many different varieties are cultivated locally, making this one of the most important crops in the region, especially in arid and semi-arid areas where cultivation is difficult.

This grass is the ancestor of cultivated millet. The narrow, loose spikes have been selected for breeding into many varieties.

Leaves are tough and somewhat wiry.

As a pioneer species, wild Finger Millet helps protect bare ground; it establishes easily, even after minimal rainfall.

6–9mm

spikelet

In the wild form, seen here, the spikelets are shed as they ripen. They are often eaten by birds or rodents.

Indian Goose Grass

Eleusine indica   Height 15–85cm

Description: A robust, vigorous and tough annual grass with leaves packed closely together in very dense clumps. Leaves are 5–35cm long, narrow and folded. The inflorescence consists of a number of spikes arranged in a cluster, usually with a few spreading outwards below the main upper cluster. Oval-shaped spikelets are neatly arranged along the length of each spike, and are variable in colour, dark green or black when young, and fading as they ripen.

Distribution & Habitat: A common and widespread grass throughout East Africa, found in a wide range of habitats and soil types. It thrives in fallow areas, on abandoned land and on roadside verges. Indian Goose Grass has a number of subspecies that occur in different habitats, from sea level to areas over 2,000m.

Ecology & Uses: A widespread and important grazing grass for both livestock and wildlife, despite the tough leaves. It can yield hay if managed when young, and it is grazed by buffalo in higher-altitude zones. The seeds are part of the diet of many different birds and rodents. This pioneer species is an important soil stabiliser.

The dark, blackish spikes emerge from dense clumps that are fringed with narrow leaves.

Inflorescences are often partly hidden among the leaves.

4.5–8mm

spikelet

These leaves have been cropped by grazing animals, but they readily sprout again to form thick, wiry clumps.

Clumps contain both young and older leaves, which provide shelter for numerous creatures.

Silvery Bottlebrush Grass

Enneapogon desvauxii   Height 5–70cm

Description: A medium-sized, clumped and compact perennial grass, often with hairy stems. Leaves are up to 12cm long, folded. The pretty inflorescence has bright silvery green spikelets arranged on an open, waving panicle that is about 10cm long.

Spikelets rapidly fall off the stems as they ripen.

Distribution & Habitat: A widespread grass of the drylands and semi-arid rangelands of East Africa, also spreading across the Greater Horn of Africa. It typically grows in wooded acacia and open semidesert grasslands, doing well on stony or sandy soils, and is often found growing with Sehima Needlegrass (*Sehima nervosum*). It is found from 400–2,000m in altitude. It also grows in the Middle East and India.

Ecology & Uses: Moderately important grazing for livestock and wildlife in arid and semi-arid areas; it is often grazed after other species have been consumed or are drying out. It forms part of the diet of various gazelles and the endangered Grevy's zebra. Because this grass can survive on poor soils in arid conditions, it is a very important part of the dryland plant community, helping to stabilise soils after rains.

Young inflorescences are paler in colour, darkening as they mature.

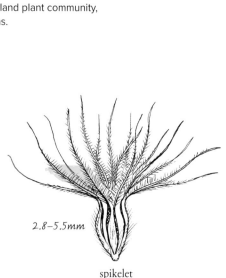

2.8–5.5mm

spikelet

One of the most striking grasses of the drylands of East Africa; stands of Silvery Bottlebrush Grass wave gently in the breeze and shimmer in the sunlight when young.

Needlegrass

Enteropogon macrostachyus   Height 80–100cm

Description: A vigorous perennial grass that grows in a loose, branched fashion, with stems growing up to a metre high. Leaves are 10–60cm long, flat or folded, with a tapered shape. The inflorescence is an elegant, narrow spike, typically single but occasionally paired; it may be erect or slightly curved. The spikelets are neatly arranged and tightly compacted. They bear short awns, initially closely pressed together in young plants and spreading out as the grass matures.

Distribution & Habitat: A widely distributed grass in East Africa and across Africa's tropical regions. It is typically found in acacia woodland and savanna grassland, including in semi-arid regions. Needlegrass grows in stands and may be interspersed with other grasses, such as Saw-toothed and Curly-leaf love grasses (*Eragrostis* spp.). This species thrives on a number of soil types, including seasonally waterlogged clays. It is found from close to sea level up to 2,000m.

Ecology & Uses: Useful grazing for both livestock and wildlife because of its leafy clumps. This grass sprouts early and grows fast after rains. The spikelets are an attractive food source for various seed-eating birds. Needlegrass has great potential if it is managed in rangelands, because it is a good grass for soil restoration and fodder production.

Reddish-purple awns are arranged around neatly packed spikelets.

7–10mm

spikelet

As soon as they sprout after rains, the young leaves are eaten by herbivores.

Pale yellow anthers dangle from the spikelets and release pollen at the slightest movement.

Tolerant of a wide range of conditions, this grass is now being developed for pasture and soil restoration under the name Bush Rye – a misnomer since it is not a rye grass.

Feather Love Grass

Eragrostis amabilis
(= Eragrostis tenella)

Height 6–50cm

Description: An elegant, often compact annual grass that grows either in an erect or a fanned-out, prostrate manner. Leaves are 7–9cm long, flat and narrow. The pretty inflorescence is a symmetrical open panicle, 2–15cm long, with spikelets arranged along branches held out from the main stem. The spikelets are greenish or purplish when young, ripening to a pale light brown colour. This species is also widely known as *Eragrostis tenella* in East Africa, but recent taxonomic work has shown it to be synonymous with *Eragrostis amabilis*.

Distribution & Habitat: A common and widespread grass in East Africa. It is found on recently bared ground, along roadside verges and well-trodden paths, and in fallow or recently cultivated areas. It grows on a number of different soils, including in sandy and rocky areas, and on soils with coral rock underlying the topsoil. Found from sea level up to 1,300m. Feather Love Grass is a pantropical species that is also found around the world.

Ecology & Uses: Seasonally grazed in varying quantities by both livestock and wildlife. It is a locally important grazing source for livestock that are tethered to graze on roadside verges or near human settlements. The seeds have been harvested for food in some regions. Feather Love Grass is also an important pioneer species that provides ground cover.

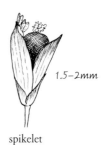

The panicle is a neat arrangement of spikelets that turn brownish as they ripen.

The inflorescence is a symmetrical open panicle.

1.5–2mm

spikelet

Highly variable in size and form, this grass provides ground cover by establishing easily on bare soils.

Woolly Love Grass

Eragrostis ciliaris Height 20–60cm

Description: An attractive, wiry perennial grass that grows singly in loose clumps or forms stands interspersed with other grasses. It often grows taller and more arched in shaded, wetter areas. Leaves are 10–12cm long, thin and narrow. The inflorescence is a dense yet graceful feather-duster-like panicle, with many short branches arranged close together along its length. It is a pretty pinkish hue when young, fading to a pale creamy colour as it ripens. The individual spikelets are tiny and closely packed in the typical overlapping form of this genus.

Distribution & Habitat: A very widespread and common grass across the drylands, desert and semidesert regions of East Africa. It often grows along roadsides and on stony hillsides, open plains and recently cleared ground, readily establishing on eroded areas. It is found from sea level up to 1,500m. This grass also grows further afield, in the deserts and drylands of the Horn of Africa, the Middle East and throughout the tropics.

Ecology & Uses: Grazed by both livestock and wildlife, this tough, wiry grass can survive in dry conditions, offering grazing in times of drought or when other grasses have been grazed down. This pioneer species readily establishes on eroded ground and quickly provides ground cover in areas that are vulnerable to soil erosion. It has potential to be managed for rangeland restoration in arid and semi-arid areas.

The feather-duster-like inflorescence changes from pink to straw coloured, adding colour to the landscape in drylands.

Flowers with tiny pinkish anthers appear early in the growing season.

2–4.5mm

spikelet

As a fast-growing pioneer species, this grass rapidly covers disturbed ground, as seen in this coastal plantation.

Button Love Grass

Eragrostis exasperata   Height 15–70cm

Other name: Bauble Love Grass

Description: An elegant annual grass that has a tufted form and grows in small, discrete clumps. Leaves are 5–20cm long, flat and narrow. The inflorescence is very pretty, almost decorative, and consists of an open-branched panicle, 7–25cm long. Individual spikelets are spread out along the length of the fine, slender stalks. The heart-shaped spikelets are often held drooping, and rustle gently in the slightest breeze.

Distribution & Habitat: A common and widespread grass that occurs in most of East Africa, and across the central and other tropical parts of Africa. This species grows along streams and on seasonally flooded rocky pans and wetlands. It grows in areas from 300–2,500m.

Ecology & Uses: An important species in the localised habitats where it occurs. Not typically grazed by livestock or wildlife, it is browsed during times of drought when other grasses are scarce. Seeds are harvested by seed-eating birds and harvester ants (*Messor* spp.). This annual rapidly establishes itself in areas that are vulnerable to erosion and trampling.

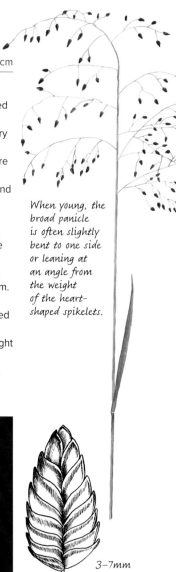

When young, the broad panicle is often slightly bent to one side or leaning at an angle from the weight of the heart-shaped spikelets.

3–7mm

spikelet

The delicate, branched inflorescence bears grey-green spikelets on fine, narrow stems.

As they ripen, spikelets turn from greyish green to straw coloured, then drop into the soil.

Stands of this grass grow in flowing water and on stream edges, providing stability and structure.

Hairy Love Grass

Eragrostis hispida   Height 10–60cm

Description: A short and compact tufted perennial grass that typically grows up to 20cm tall. Leaves are 15–20cm long, extremely fine, covered in hairs, and with delicate pointed tips. The inflorescence is very distinctive and pretty, consisting of an open panicle with spikelets clustered in groups arranged at intervals ascending the spike, somewhat like Christmas decorations. Individual spikelets are covered in stiff, densely packed bristles that give a silvery, hairy appearance.

Distribution & Habitat: A grass that grows on damp, shallow soils in areas with underlying rocks. It is often found on the edges of escarpments, where moisture and rain seep through. Hairy Love Grass grows in association with rock violets (*Craterostigma* spp.) and Tiny Dropseed Grass (*Sporobolus festivus*). The species is found from 1,000–2,500m in altitude.

Ecology & Uses: Not heavily grazed by livestock or wildlife. The fine leaves are gathered by birds to build and line nests. This grass is an important stabiliser for thin, fragile soils in the microhabitats that it occupies. It can be used in dried-flower arrangements, as the silvery spikelets retain their shape and colour for long periods once dry.

Hairy Love Grass is found in association with many other plant species, creating a unique and attractive microhabitat.

2.4–6mm

spikelet

In this aptly named species, the stems, leaves and spikelets are covered in dense, fine hairs.

As the grass ripens, the delicate inflorescence sheds spikelets that fall onto the damp soil below; the seeds inside can wait many years for rain before they germinate.

Streamside Love Grass

Eragrostis namaquensis   Height 10–150cm

Description: A variable, elegant tufted annual grass that typically grows in a curved fashion. Leaves are flat and up to 30cm long. The arched inflorescence is 20–30cm long. Spikelets are tiny, clustered closely together along the drooping stems, which are a dark green colour when young. They easily break off when the plant is handled. As they ripen, the spikelets fall away, leaving a wispy, bare, arched inflorescence.

Distribution & Habitat: A widespread species, typically found at the edges of seasonal pools and streams. It often grows among rocks on streambanks, interspersed with other grasses, including water grasses (*Echinochloa* spp.). Found from 300–2,000m in altitude.

Ecology & Uses: This species grows in a restricted habitat. It is not heavily grazed by livestock or wildlife, apart from during droughts when animals gather more frequently at the edges of streams in search of food. Weaverbirds use the leaves and stems for nest building, and seed-eating birds consume the spikelets.

A grass that is harder to spot when young, as the narrow inflorescence is often hidden among other grasses and plants.

Streamside Love Grass grows hidden among other grasses along rocky seasonal streams and wetlands.

1.3–2.2mm

spikelet

Typical rocky seasonal-stream habitat

Dryland Love Grass

Eragrostis racemosa   Height 5–10cm

Description: A typically short, tufted perennial grass that grows in a low, clumped fashion. Leaves are 5–10cm long, flat. They often curl into a rolled shape when dry. The inflorescence is an open panicle along which spikelets are elegantly arranged. Spikelets are a symmetrical, tapered oval shape; the edges are complete, not jagged or serrate. Pale green when young, the spikelets quickly turn to a pale, almost ivory colour as the plant matures.

Distribution & Habitat: A widely distributed species across the drylands and savannas of Africa. In East Africa this species is more typical of drier areas, especially northern Kenya and Uganda. It typically grows on shallow soils and can do well on stony ground. Found from 300–2,400m in altitude.

Ecology & Uses: As a fast-growing species that is leafy when young, it is grazed by both cattle and sheep, mostly the latter, because its short leaves are readily within reach. It is also an important dryland grazing species for wildlife, including the endangered Grevy's zebra, the common zebra, and various subspecies of hartebeest.

Spikelets are densely arranged in variable groups of 5–16.

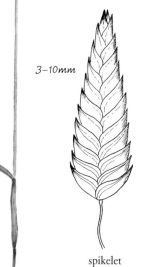

3–10mm

spikelet

Prostrate clumps form on bare soils in hot, dry areas.

During the brief flowering stage, tiny white anthers emerge from the gaps between the spikelet parts.

Curly-leaf Love Grass

Eragrostis rigidior   Height 30–100cm

Description: A pretty, erect grass that grows in dense stands, sometimes interspersed with other grass species. Leaves are 10–25cm long, flat and fairly stiff when young; older leaves persist on the plant, forming curly strands as they dry, which distinguishes this species from the other love grasses (*Eragrostis* spp.). The inflorescence is a symmetrical panicle that stands elegantly above the leaves; it opens out in a whorl at the base, with grey-green spikelets on branches that rise in an alternating fashion to the tip of the panicle.

Distribution & Habitat: A widely distributed grass in East and southern Africa. It grows on a range of soil types, but in East Africa it is typically found on red loam soils. It is a tough grass that can survive on rocky ground and even grow on overgrazed land. It often grows in recently fallowed areas following rains. Found 1,000–2,500m above sea level.

Ecology & Uses: A fast-growing species that is quickly established, even after poor rains, making it an important grazing species in semi-arid areas. Typically grazed by livestock when it is young, as the leaves are more tender. This species is also grazed by wildlife such as the endangered Grevy's zebra, the common zebra, elephant and various species of gazelle. The seeds are eaten by different birds, including waxbills, and by rodents such as striped grass mice.

Older leaves form a distinctive curled shape as they age, giving this plant its common name.

The broad, open panicle has many spikelets neatly arranged along its branches.

3.5–7mm

spikelet

Young leaves are flat and stiff.

Dense stands appear early in the rainy season; here growing with Needlegrass.

Tall, open panicles grow and ripen above the leaves after good rains; inflorescences are stunted when rainfall is poor.

Curly-leaf Love Grass ■ 125

Highland Love Grass

Eragrostis schweinfurthii   Height 10–50cm

Description: An elegant, often petite annual grass with a compact growth form. Leaves are 1–8cm long, flat with a short, tapering point. The inflorescence is an attractive, symmetrical panicle with the spreading branches decreasing in size as they ascend the stem, forming an elongated triangle. Individual spikelets are 4–10mm long, with the classic overlapping and tapered shape of the genus; they are grey-green when young and fade as they ripen.

Distribution & Habitat: A common and widespread grass in the highland regions of East Africa. It typically grows on soils with underlying rock, on roadside verges, in ditches, and occasionally in forest glades. The species is mostly found from 300–1,600m above sea level. It also grows across tropical Africa and parts of Asia.

Ecology & Uses: Moderately grazed by wildlife, including various species of forest antelope. This grass is an important member of the grass community that helps to provide ground cover and stabilise soil in highland areas.

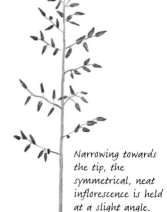

Narrowing towards the tip, the symmetrical, neat inflorescence is held at a slight angle.

4–10mm

spikelet

Spikelets have a purplish tinge when the grass grows in exposed areas at higher altitudes; this is a response to more intense sunlight.

Highland Love Grass often establishes in ditches and recently cleared areas, where it provides ground cover and shelter for insects and other creatures.

Dwarf Love Grass

Eragrostis sennii
(= *E. abrumpens*) Height 10–20cm

Description: A pretty, compact annual grass that typically grows in small, neat clumps or partially prostrate, close to the ground. Leaves are 7–15cm long, soft and flat, or slightly rolled. The inflorescence is a neat panicle, with spikelets arranged in an elegant fashion along the main stem. Spikelets are a narrow, oval shape and very pretty, with the fine overlapping pattern and serrated edges clearly visible, 8–18mm long. They are a bright pale green when young, ripening to a pale straw colour before they readily drop off the plant.

Distribution & Habitat: A common, widespread grass in coastal and near-coastal regions of East Africa and the Horn of Africa. Typically found on roadside verges and along paths and livestock tracks in open areas, and where the ground is subject to a great deal of trampling or traffic. It grows and thrives on sandy and rocky soils. Found from sea level up to 100m.

Ecology & Uses: Moderately important grazing grass, seasonally grazed by both livestock and wildlife in all coastal regions of East Africa, especially in mixed-farming systems and where livestock is kept tethered on roadside verges or in communal land during the day. It is one of the grasses grazed by the hirola, a rare and endangered antelope in Kenya. Because Dwarf Love Grass can survive on bare ground in hot, dry conditions, it is an important pioneer species, providing ground cover.

One of the prettiest of the love grasses; this compact, dainty plant is much tougher than it looks.

Tiny anthers are just visible for a short time during early stages of growth.

8–18mm

spikelet

This grass readily establishes on bare, rocky and sandy ground, even after brief rains, providing cover and structure for other species. Here it is growing alongside weedy *Euphorbia hirta* (sprawling on runners, with dark green and purplish rounded leaves).

Saw-toothed Love Grass

Eragrostis superba　　　　　　　　Height 50–100cm

Other name: Sawtooth Love Grass

Description: One of the most beautiful and recognisable of African grasses, this tufted, erect perennial is a medium-sized grass that grows in both stands and clumps. The leaves are flat, 10–40cm long. The inflorescence consists of an open panicle. Large, heart-shaped spikelets have distinctive saw-toothed edges and are arranged in an alternating pattern, ascending to the tip of the panicle. They are a striking, bright reddish colour on young plants until the peak of growth, but they turn pale as they dry, eventually dropping to the ground, where they collect under and around the mother plant.

Distribution & Habitat: One of East Africa's most widespread grasses. It is found in a range of habitats, primarily in mid-altitude regions in open grassland, savanna and bushland. This species can grow on a wide range of soils, including those in seasonally wet or flooded areas. Found from 500–2,500m in altitude.

Ecology & Uses: An important grass for grazing, both by livestock and wildlife. It can be managed for fodder and baled as hay. Many different birds and insects, including harvester ants (*Messor* spp.), use the seeds as food. This fast-growing species has juicy leaves and establishes quickly following rains. The large seeds are easy to collect and store. This grass has been used in rangeland restoration efforts in a number of dryland areas.

Spikelets are large and easily collected and stored for rangeland restoration efforts.

6–16mm

spikelet

Spikelets are clustered closer to the stem in younger inflorescences.

Maturing spikelets spread out to form a graceful panicle as they ripen.

This is one of the region's most abundant grasses – a delight to see after good rains.

Weedy Love Grass

Eragrostis tenuifolia   Height 10–80cm

Description: A medium-sized, clump-forming grass that typically grows in a straggly fashion or partly sprawls along the ground. It has hard and wiry leaves and stems that are difficult to pull out of the ground. Leaves are flat and narrow, 4–30cm long. The inflorescence is a panicle sitting above the leaves, with narrow, pointed spikelets that are grey-green when young, turning straw coloured with age and opening up slightly, giving the spikelet edges a jagged appearance.

Distribution & Habitat: A very common, widespread grass found throughout East Africa and most other tropical regions of the world. It typically grows at the edges of roads and paths, and on disturbed ground. The species can grow in partly shaded areas too, and will form stands in areas that have been left to fallow in more moist or humid habitats. Found from sea level up to areas 3,000m in altitude.

Ecology & Uses: Not a heavily grazed grass, especially once it matures, owing to its tough, wiry nature. Its seeds are eaten by granivorous birds. This grass is one of the weedy pioneers that rapidly colonise bare ground, making it an important species for soil stabilisation and creating a structure in which other plant species can become established.

Young panicles have many tiny spikelets that drop off as they ripen, leaving behind empty, dry stems.

4–16mm

spikelet

During wetter years, this grass forms larger, more open panicles, resembling those of the Curly-leaf Love Grass.

Weedy Love Grass is one of East Africa's most widespread and variable grasses.

This grass is a very important stabiliser of bare or disturbed ground.

Tussock grasses

Festuca species

Height 30–75cm

Description: Several species of tussock grass are similar in growth form and habitat. These perennials grow in distinct, dense, round clumps shaped like large spherical pom-poms. Fine, narrow leaves are extremely tough, 10–30cm long. The inflorescence is a narrow, fine panicle borne on a stem that is held up above the dense, leafy clumps. Spikelets are an elongated oval shape, green when young and fading to cream, often persisting in a dry form once they have ripened and the seeds dispersed.

Distribution & Habitat: Tussock grasses are found on all the high mountains of East Africa, where they are a classic feature of the moorland or Afro-alpine vegetation zone. They occur from 2,000m up to sheltered areas above 4,250m on Mounts Kenya, Elgon and Kilimanjaro, as well as on other high mountain ranges in the region.

Ecology & Uses: With their tough, wiry leaves the tussock grasses are not typically grazed much by livestock or wildlife. The older leaves persist, forming a dense, packed clump, which helps to protect the plant during freezing nights. These grasses are very important within the high-altitude habitats, providing structure for other plants, and refuge for many different rodents, montane lizards and snakes. Their seeds are eaten by birds. Individual plants are long lived and slow growing, persisting for many decades.

Spikelets are a green to purplish colour when young, fading as they ripen.

Inflorescences rise on a long, narrow stem from a dense clump. Abyssinian Tussock Grass (illustrated here) is widespread in high-altitude areas of East Africa.

6–12mm

spikelet

Tussock grasses form large clumps in high-altitude Afro-alpine environments. This habitat is characterised by unique plants such as the flowering giant lobelia (*Lobelia deckenii*) in this grassland on Mount Kenya.

Short Spear Grass

Harpachne schimperi Height 15–30cm

Description: A fairly short perennial grass that grows in dense tufts. Leaves are flat, up to 19cm long, with a fine, pointed tip. The distinctive inflorescence ensures that this grass cannot be mistaken for any other species. Spikelets are densely packed together on the raceme, pointing downward at an angle from the main stem. Each spikelet has a serrated outline, is green and tinged with pink and purple when young but fading to a pale straw colour as the plant ages.

Distribution & Habitat: Widely distributed from mid-altitude to highland areas across East Africa. It grows in open grassland and in glades in bushland habitats. The species can grow well on rocky or sandy soils, often forming dense stands in patches covering several square metres. It is found in areas 500–2,800m in altitude.

Ecology & Uses: A pioneer grass foraged by livestock and wildlife in dryland areas. It can survive where conditions are often quite harsh, making it a useful species, even if the overall leaf biomass is low. Birds, harvester ants (*Messor* spp.), rodents and some mammals, including the enigmatic spring hare, make use of the seeds and young stems for food. This grass is familiar to anyone who has walked through grassland in East Africa, owing to the seeds rapidly embedding in fur or footwear.

The inflorescence is typically ankle height – perfectly placed for spikelets to cling to socks or furry limbs for seed dispersal.

8–21mm

spikelet

Younger inflorescences are light green and tinged with pink.

This grass can grow in rocky areas and does well even on poor-quality soils.

A field of ripening Short Spear Grass is the hiker's bane as the spikelets will catch onto socks and shoes in large numbers!

Tall Spear Grass

Heteropogon contortus   Height 30–100cm

Description: A tufted, leafy perennial grass that forms light spreading clumps, with leaves and stems a striking bright green when young, fading to a yellowish colour as the plant matures. Leaves are 3–30cm long. The inflorescence and spikelets are distinctive; each narrow spike has individual spikelets arranged in a closely pressed pattern, packed together. As these mature, their long, twisted awns become entangled with each other, forming untidy massed clusters. The awns darken as they age and ripen.

Distribution & Habitat: A common and widespread grass found in a wide range of habitats, including woodland, wooded grassland, bushland and open grassland; also growing among rocks and along the roadside, colonising eroded ground. Although this species thrives in coastal glades and on dunes, it occurs up to 2,100m above sea level.

Ecology & Uses: An important grazing grass when young, the relatively nutritious leaves are favoured by larger herbivore species that live in more wooded habitats, including sable and roan antelopes. Hikers will know the sharp-pointed seeds well, as they embed in socks and shoes, and can give a painful prick when least expected (hence the common name). The long awn and pointed tip equip the seeds to partly bury themselves, which enables them to survive drought, fire and overgrazing.

As the inflorescence ripens, the awns – which can be up to 80mm long – become increasingly twisted, often intertwining with each other and neighbouring grasses.

Spikelets bear long, slender awns, straight and packed together when young.

Many different insects, like this chafer beetle, are lured by the young inflorescences.

5.5–10mm

spikelet

A stand of young Tall Spear Grass in a rocky area; this grass can be spotted almost anywhere in East Africa, from the coast to the savannas.

Common Thatching Grass

Hyparrhenia hirta Height 30–100cm

Description: A tall, striking grass that is part of a diverse group of thatching grasses. It grows in dense or loose stands, gracefully erect, waving and shimmering in the sunlight as the plants mature. Leaves are narrow, with a slightly rough feel to the touch, 2–15cm long. The inflorescence is an elongate, loose panicle with the individual spikelets typically facing to one side, often in a slightly drooping manner. The awns project from the spikelets, often slightly entwined or curling as they age and ripen. The spikelets and bases are covered with silvery hairs that become more visible as the plant ripens.

Distribution & Habitat: A common and widespread grass in East Africa. Found more commonly in the highland areas, on open dry hillsides, roadside verges and gentle rocky slopes, and at the edges of forest or woodlands. It grows at altitudes from 1,300–2,700m above sea level.

Ecology & Uses: As the common name suggests, this species, together with other species of *Hyparrhenia*, is widely used as a thatching grass for houses, outdoor buildings and granaries, making it highly valued in certain areas. Thatched houses are cooler in hot weather, and a well-thatched roof can last for many years, even in the tropical climates of East Africa. This grass is also grazed by livestock and wildlife when it is young and in fresh leaf.

Awns are 10–35mm long and help the grass to disperse its seeds.

Younger inflorescences are bright green with hints of copper.

4–6.5mm

spikelet

The silvery hairs become more visible as this grass ripens and glistens in sunlight.

Fields of Common Thatching Grass provide shelter for many different grassland birds and other animals.

Yellow Thatching Grass

Hyperthelia dissoluta   Height 90–300cm

Description: A tall, striking perennial grass that grows in an erect, arched fashion, occasionally reaching up to 3.5m. The stems are a bright yellow, easily seen from a distance, with bright green sheathing leaves. Leaves are narrow, 10–30cm long. The inflorescence is a pretty, open panicle that takes up about one-third of the total length of the plant, often leaning slightly to one side. Spikelets are arranged in elegant pairs emerging from a short sheath, with distinctively long ochre-yellow awns that extend up to 10cm. The awns are straight when young, becoming more twisted and spiral-shaped as they age and ripen. They attach to passing animals or snag on vegetation, so enabling the dispersal of the seeds.

Distribution & Habitat: A common and widespread grass in East Africa, found in a wide range of habitats, including open woodland, bushland, tall grassland and fallow land. This species thrives on disturbed ground and is often found sprouting after fires. It grows from sea level up to areas 2,500m in altitude.

Ecology & Uses: When young, this grass is an important source of grazing for both livestock and wildlife, but is grazed less as it matures. Widely harvested in many areas for use in thatching roofs and making mats. It is also used by birds and insects for shelter. The species is managed for hay and silage production.

The complex inflorescences can be spotted from a distance, making this an easily identified species.

Awns become more curled as they ripen.

9–15mm

spikelet

Insects feed and shelter on Yellow Thatching Grass. Here a grass stink bug uses its piercing mouthparts to suck sap from a young spikelet.

Young spikelets bear long, straight awns that are yellow to copper in colour.

Sisal plantations and fallow grounds are some of the areas where this grass grows very successfully.

Red Lintonia

Lintonia nutans   Height 20–80cm

The inflorescence comprises several drooping racemes.

Description: An elegant, medium-sized perennial grass with a robust growth pattern that sprouts in tufts from short rhizomes. Leaves, up to 22cm long, are flat and narrow, an attractive pale bluish-green in colour. The inflorescence consists of 2–4 (sometimes up to 6) pretty, curved and drooping racemes. Rounded spikelets are arranged in an alternating pattern, pressed against the stems that bear them. They are a deep red-purple colour when young, fading as the plant ages.

Distribution & Habitat: This is a fairly widespread grass in East Africa. It grows on black-cotton clay soils, where it thrives in the seasonally waterlogged conditions. It is often found in association with Hairy Signal Grass (*Brachiaria lachnantha*) and wire grasses (*Pennisetum* spp.). Found at a wide range of altitudes, from sea level to over 2,000m.

Ecology & Uses: A grass of moderate grazing value for livestock and wildlife. It is grazed by zebra and cattle in the dry season, as they move into the clay soil grasslands in search of pasture. The heavy, rounded spikelets are much sought after seasonally by seed-eating birds, including waxbills, finches and quelea.

The red inflorescence is easily spotted as the grass matures.

5–11mm

spikelet

This grass forms loose clumps that are typically interspersed with other grasses in areas of black-cotton clay soils.

Sour Grass

Loudetia simplex　　　　　　　Height 30–150cm

Description: An attractive perennial grass with an upright growth form. The sheaths that cover the bases of the older stems often become frayed, giving the clumps a silky appearance. Leaves are narrow, flat or slightly rounded, 10–30cm long and only 2–5mm wide. The inflorescence is a graceful, loose panicle that waves in the softest breeze. The spikelets are 8–13mm long, ranging in colour from light to dark brown, and bearing awns that range from 2.5–5cm long; these often become more curved or twisted as they ripen.

Distribution & Habitat: A common and widespread grass throughout East Africa. Typically it is found growing in small stands on rocky hillsides, and in open grassland and areas that are seasonally wet. This species generally occurs 1,000–2,600m above sea level.

Ecology & Uses: A grass that is not typically grazed, owing to its dense, wiry nature. However, in times of drought or reduced rainfall, it is grazed by both livestock and wildlife. Sour Grass grows in association with other specialised grasses and wildflowers. It recovers more quickly from fires, which makes it an important ground cover that helps areas to regenerate, subsequently preventing soil erosion.

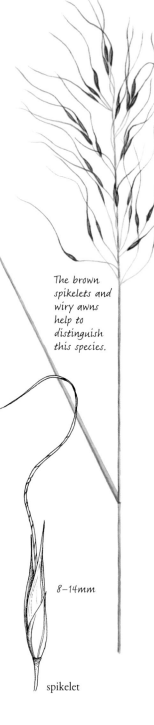

The brown spikelets and wiry awns help to distinguish this species.

8–14mm

spikelet

This grass can thrive even on a tiny patch of soil or in a slim rocky crevice.

Sour Grass has a more compact appearance when young as the spikelets and awns are clustered together.

It opens into a graceful panicle as it matures; long awns are clearly visible.

Clumps of this grass form an important shelter for insects, birds and smaller creatures in rocky areas.

Forest Megastachya

Megastachya mucronata Height 10–100cm

Description: An attractive, distinctively broad-leafed, tufted annual grass that grows in a slightly spreading fashion. Wide leaves have an elegant, wavy margin tapering to a narrow tip, 10–15cm long. The inflorescence is a distinctive, beautiful open panicle, 15–25cm long, with spikelets arranged around it on short, thin stalks. Spikelets are a narrow, oval shape, 7–14mm long, bright green when young and fading to pale creamy brown as the plant ages.

Distribution & Habitat: A common and widespread grass in the warmer, wetter regions of East Africa. It typically grows in the shade along forest and woodland paths and roads. Forest Megastachya is found in areas from sea level up to 1,000m in altitude.

Ecology & Uses: An important grass that thrives in forest shade, forest edges and woodland habitats. It is moderately grazed by both livestock and wildlife, and browsed by grazers and other herbivores, such as common duiker and bushbuck, that live in shady habitats. Many insects use this grass for shelter, including male *Amegilla* bees that gather to sleep on it in small groups.

Young spikelets are bright green and neatly pressed together on skinny stalks.

Leaves have a wavy margin.

Many insects take shelter on this grass, like these male *Amegilla* bees sleeping on the stem of the inflorescence.

7–14mm

spikelet

Both young (green) and mature (pale straw colour) inflorescences are often found side by side in the dappled shade of coastal forests and woodlands.

False Guinea Grass

Megathyrsus infestus
(= Panicum infestum)

Height 50–200cm

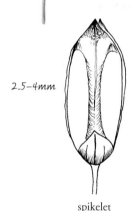

Description: An elegant and robust, tufted grass that forms large, dense clumps and grows from a short rhizome. Leaves are 15–50cm long and flat, with a fine, tapering tip. The inflorescence consists of a graceful, open panicle that is loosely branched, 10–28cm long. Each branch consists of individual spikes held out at an angle from the central axis and often slightly curved. Spikelets are grooved and oval shaped, pale yellow-ochre to greenish in colour when young, becoming paler as they ripen.

The loosely branched, open inflorescence gives this grass an elegant appearance.

Distribution & Habitat: A common and widespread grass in the coastal, lakeside and low- to mid-altitude areas of East Africa. It is also found throughout dryland and tropical regions of Africa. This species grows along the edges of forests, in open woodland, bushland and tall-grass coastal glades, and on roadside verges and abandoned fallow land. It grows from sea level up to areas 1,500m in altitude.

Ecology & Uses: As a widespread species, it is of value to both wildlife and livestock. In some habitats, such as coastal glades in the Arabuko Sokoke Forest Reserve in Kenya, it is an important grazing species for elephant and buffalo. It is seasonally harvested for fodder in some areas, especially when other grazing is scarce. Numerous birds and insects make use of the stems, seeds and leaves for nesting and food.

2.5–4mm

spikelet

When in flower, the grass has brownish anthers that extrude from the spikelets.

Tall stands of this grass can often be found at the edges of open glades, fallow fields and roadside verges.

Red Top Grass

Melinis repens Height 30–100cm

Other name: Shining Top Grass

Description: One of the most recognisable African grasses, this pretty perennial grass grows in stands, forming dense mats from a creeping base, typically rising up to 50–75cm. Bright pale green leaves are flat or folded, with a fine, pointed tip. The inflorescence is a distinctive panicle that is a bright rosy pink when young, and ages to a pale silvery colour. Spikelets are covered in dense, soft hairs.

Distribution & Habitat: A very widespread and common species across East Africa. The grass is often found on roadside verges, recently disturbed or fallow land, and rocky hillsides. It grows and thrives at mid-altitude areas, generally 900–2,500m above sea level.

Ecology & Uses: Moderately grazed by both livestock and wildlife, but more heavily grazed when young. The dense, mat-like growth makes it an important fodder species in drylands. The seeds are eaten by birds and widely dispersed by mammals, which carry it elsewhere on their coats and skin. Red Top Grass is an important pioneer species for restoration of bare ground. Stands of this grass, catching the sunlight in the morning or evening, are one of the most beautiful sights to be had in the region.

The inflorescence is a distinctive panicle with softly hairy spikelets.

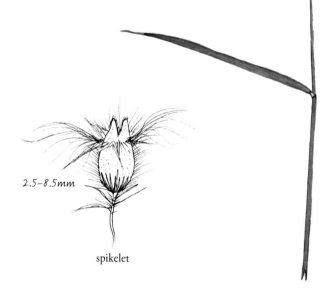

2.5–8.5mm

spikelet

Younger plants have more reddish inflorescences that turn silvery as they ripen.

Fields of Red Top Grass glisten in the evening or morning sunshine, giving the landscape an air of shimmering beauty.

Pincushion Grass

Microchloa kunthii   Height 10–60cm

The inflorescence is a delicate, curved spike.

Description: A short, compact perennial grass that grows in dense mat-forming tufts, typically 10–20cm long. Leaves are 1–8cm long and very fine, clustered together at the base of the plant. The inflorescence consists of a long, fine, slightly curved spike, reaching straight up from ground level. Spikes are a dark reddish brown on young, fresh plants, fading to light brown as they age.

Distribution & Habitat: A widely distributed grass across East Africa. It is typically found growing on thin, shallow soils on top of rocky pans, among rocky outcrops and in rocky soils that are seasonally waterlogged. This grass is often found in association with Hairy Love Grass (*Eragrostris hispida*) and Tiny Dropseed Grass (*Sporobolus festivus*).

Ecology & Uses: Given the distribution of this grass and the rocky microhabitat that it occupies, it is not an important species for grazing by livestock or wildlife. Because it grows and thrives in rocky, seasonally wet areas, it plays an important role of providing structure in a diverse community of specialised wildflowers and insects that are found in these habitats.

2.5–4mm

spikelet

When young, this grass can be harder to spot, but it is common on rocky pans.

The fine, short leaves help the plant to survive in areas that dry up quickly.

At the flowering stage, the narrow, slightly curved inflorescences bear rows of purple anthers, which extrude from the dark spikelets.

Forest Basket Grass

Oplismenus burmannii   Height 5–60cm

Description: A straggly perennial grass that grows in a trailing fashion, often draped or creeping over other plants and the ground. Leaves are 1–10cm long and a strong dark green, with a distinctive wavy margin. The inflorescence consists of a series of short, symmetrically arranged spikes along the stem, held above the plants' leaves; spikes are 2–10cm long. The individual oval-shaped spikelets, 2.5–4mm long, are arranged in a row on the stalks, closely packed together, covered with fine hairs. These bear distinctive fine awns, up to 15mm long, that feel slightly sticky and adhere when touched.

Distribution & Habitat: A common and widespread grass in East Africa. It thrives in shade or dappled light conditions, sometimes forming small clumps around the roots of trees and along forest paths. Forest Basket Grass grows in forest and woodland areas, and along streams in riverine forest, from sea level up to areas over 2,100m in altitude. This species has been introduced to many other parts of the world. In some areas it is considered invasive.

Ecology & Uses: This species is moderately grazed by forest herbivores and rodents. Because it grows in wetter areas and can survive in shaded, damp forest conditions, it is an important part of the forest floor ecology, providing ground cover and preventing soil erosion, especially along forest streams, paths and trails in high-rainfall areas.

Fine awns feel rough to the touch and adhere easily to fur or skin.

Fine, whiskery awns extend from closely arranged spikelets.

2.5–3.5mm

spikelet

As this grass can establish along paths and trails, it is an important species for stabilising soil and preventing erosion.

This grass forms short, dense mats where there is sufficient rainfall, creating ground cover that provides shelter for different animals and protects soils.

Basket Grass

Oplismenus hirtellus   Height 15–100cm

Description: A variable, rambling prostrate grass that typically grows in a loose fashion, draped over roots or rocks. It roots readily at the nodes. Leaves are 1–15cm long and distinctive, with an elegant tapered shape and a wavy, pattern-like structure. The inflorescence is borne 3–15cm above the plant, with short clustered groups of spikelets arranged along its length. The spikelets are 2–4mm long, with sticky reddish-purple awns.

Distribution & Habitat: A common and widespread species that grows in forests and wetter woodland areas, and along streams in riverine forest. It typically grows in shade or dappled light conditions, sometimes forming small clumps. Basket Grass is found from sea level up to areas over 2,500m in altitude. It is widespread across most of the wet tropics of the world.

Ecology & Uses: This species is moderately grazed by forest antelopes and other grass-feeding animals, and is eaten by birds that live in the forest undergrowth. As it grows in wetter areas and is one of the grasses that can survive in shaded, damp forest conditions, it is an important part of the forest floor ecology, used by many different insects and molluscs. Some varieties are grown in hanging baskets for ornamental purposes, hence the common name.

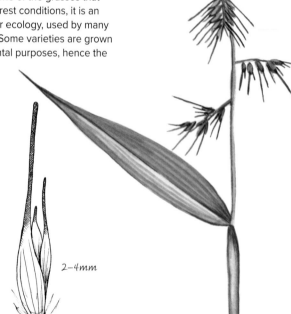

Short awns are slightly thickened; when ripe, they cling to passers-by.

The inflorescences are easily spotted when illuminated by sunrays in forest undergrowth.

2–4mm

spikelet

Basket Grass can form dense, untidy stands interspersed with short herbaceous species and wildflowers.

Red Rice Grass

Oryza punctata   Height 60–150cm

Description: A pretty, erect annual grass that typically grows in a fairly dense stand with the stems rising up in a loose, somewhat untidy fashion. The clumps are a bright lemon-green colour when seen from a distance. Leaves are flat, up to 30cm long, with tapered tips. The inflorescence is an open, loose panicle with spikelets on branches, held close to the short stems. The spikelets are distinctive, with a slightly rough surface, typically pale green with reddish markings. Often a few of the spikelets appear almost black, these readily detaching from the plant. The spikelets bear long, thin yellowish awns up to 7cm long.

Distribution & Habitat: A common and widespread grass in coastal and near-coastal regions of East Africa, and in the wetter, more tropical inland regions of western Kenya, Uganda and Tanzania. It thrives in wetlands, including on flooded land and along the edges of slow-moving streams and ponds. This species is found from sea level up to 1,000m in altitude. It is also found in Madagascar and has been introduced into parts of tropical Asia.

Ecology & Uses: An important grass in wetlands, where it provides cover, shelter and food for many different birds and insects. It is moderately grazed, especially in the dry season when other grasses are not available. The seeds are harvested and used as food in some areas, especially during droughts or famines. Red Rice Grass can sometimes be a weed in paddies where cultivated rice (*Oryza sativa* and *O. glaberrima*) is grown.

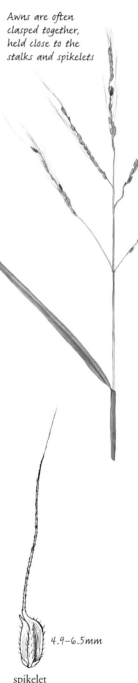

Awns are often clasped together, held close to the stalks and spikelets

Spikelets appear after a few months of good rains.

4.9–6.5mm

spikelet

The inflorescences mature gradually and are often found in a somewhat untidy arrangement.

Seasonal wetlands, like this coastal bushland, are a typical habitat for this widespread species.

Red-seed Guinea Grass

Panicum atrosanguineum   Height 10–40cm

Description: A dense, short and compact annual grass with hairy sheaths and leaf blades. Leaves are flat and narrow, 5–15cm long. The inflorescence is delicate – a large and open spreading panicle that often makes up over half of the plant's total size. Spikelets are distributed along the entire panicle, and are a dark red colour when young, eventually fading to a straw colour as the seeds ripen.

Distribution & Habitat: This grass species is widely distributed in East Africa. It typically grows on disturbed ground, recently cultivated areas and roadside verges, and is indicative of poor soils. It is often found growing in shady or partly shaded sites situated 100–2,300m above sea level.

Ecology & Uses: This grass is of reasonable grazing importance, especially seasonally in drylands and rangelands. The seeds are eaten by a number of birds, including waxbills and finches. It is a weedy species and it readily establishes on bare ground, where it helps provide ground cover and soil stability; it therefore plays a role in creating better conditions for more nutritious grasses to succeed.

The bright red inflorescence is typically seen much later in the rainy season, well after the leaves are established.

1.8–2.2mm

spikelet

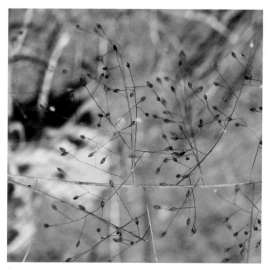

Panicles often contain both young and ripening spikelets; inflorescences persist for some time.

This grass helps to prevent soil erosion as it grows on bare ground from seeds that were previously dispersed.

Small Guinea Grass

Panicum coloratum   Height 15–100cm

Description: A dense, variable perennial grass that grows in tufted clumps, either with stems flat on the ground or draped over other grasses or vegetation. Leaves are long and flat, typically hairy, 6–30cm long. The inflorescence consists of a loose, open panicle, 4–30cm long. Spikelets are oval-shaped and typically purplish red.

Distribution & Habitat: Widely distributed throughout the East African region, typically in grassland and bushland habitats. Small Guinea Grass often grows in stands on seasonally wet soils, including black-cotton clay soils. It grows at a wide range of altitudes, from close to the coast up to areas over 2,300m in altitude.

Ecology & Uses: One of the most important grazing and fodder grasses for both livestock and wildlife throughout East and southern Africa. It has been managed for hay, and produces a good yield. This species is indicative of healthy soils and good grazing conditions. It is grazed by a wide range of herbivores, including buffalo, reedbuck and rhino, and the seeds are favoured by many different birds. The stems are host to a number of stem-boring insects such as beetles and moths.

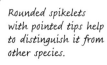

Rounded spikelets with pointed tips help to distinguish it from other species.

A field of Small Guinea Grass often needs a closer look to spot the inflorescences.

2–3mm

spikelet

When in flower, as seen here, the spikelets are dotted with bright orange-yellow anthers that shed pollen into the air.

Hairy Guinea Grass

Panicum deustum   Height 70–200cm

Description: A robust, striking perennial grass that grows upright; it expands by spreading from a short, hidden rhizome, often in a slightly untidy fashion. It rises up above other grasses and surrounding vegetation. Leaves are 15–50cm long, a bright lemon-green colour, slightly rounded at the base and narrowing towards the tip. The inflorescence is a pretty, open panicle in a loose tapering shape, 10–40cm long, arranged in a series of branches that point upwards. The individual spikelets are a stocky, oblong shape, 3.5–5mm long, pale green when young and ripening to a striking red-purple colour; these often drop off when the plant is touched or handled.

Distribution & Habitat: A common and widespread grass in East Africa. It tends to grow in slightly more moist locations, among rocks, on clay soils, at the edges of rocky outcrops and near streams. Found in areas near sea level up to over 2,000m in altitude.

Ecology & Uses: An important seasonal grazing species for both livestock and wildlife. The lush, leafy clumps are often very well grazed by elephant, buffalo and other species, sometimes so much so that it is hard to find the grass in seed because it has been heavily grazed when young. It is also used by many different birds and insects. The relatively large seeds attract a number of specialised seed-feeding insects.

Often found in areas with Tall Guinea Grass, but easily recognised by much plumper and more densely arranged spikelets

The anthers and stigma protrude from the ends of the spikelets when in flower.

The brightly coloured inflorescences are highly visible when flowering.

3.5–5mm

spikelet

Hairy Guinea Grass forms large, dense, persistent clumps that provide shelter for birds and other animals, and attract many insects.

Tall Guinea Grass

Panicum maximum   Height 1–2.5m

Other name: Buffalo Grass

Description: A typically tall, elegant and striking grass, easily recognised by its imposing inflorescence, held on a long stalk high above the leaves. The plant can reach over 2.5m tall, but is mostly around 1–2m. Leaves are straight and flat, 6–45cm long. The inflorescence is an open panicle, with spikelets loosely arranged in long whorls, decreasing in size as they reach the tip. Spikelets are bright green, tinged reddish purple when young, and falling off easily when the plant is touched or moved.

Distribution & Habitat: A very widespread species found throughout East Africa. It grows in a range of habitats and on a variety of different soil types. It is often found in open woodland and bushland, and on the edges of riverine or waterlogged areas. Tall Guinea Grass grows from sea level up to 2,400m in the highlands. The species is also widely distributed in the tropics.

Ecology & Uses: The robust growth pattern, high leaf production and high protein content of the leaves make this one of the most important fodder species for livestock. It is also widely browsed by a variety of wildlife, including elephant and Cape buffalo. The seeds are a favourite food of many different waxbills, finches and guineafowl. The thick stems host a number of stemborer moth species.

Spikelets are arranged in loose whorls.

Reddish-purple spikelets are loosely arranged on the panicle.

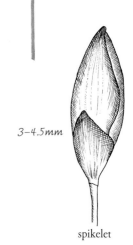

3–4.5mm

spikelet

Stalk-eyed flies are often found on the leaves of Tall Guinea Grass in wetter habitats.

A variety of crickets and grasshoppers live and feed on this grass.

Some butterflies use this as host plant for their larvae, including the common evening brown.

Dozens of different herbivores graze on this grass, including the largest land animal, the African elephant.

Ditch Millet

Paspalum scrobiculatum   Height 30–60cm

Other names: Kodo Millet, Koda Millet

Description: A short, scruffy perennial grass that often forms clumps or dense mats. Pale green leaves, 5–40cm, are soft and juicy, growing in a slightly overlapping manner. The inflorescence consists of 1–4 short spikes, typically 2, emerging from between the leaves and held at a wide angle from the main stem. Spikes are green and fleshy to the touch, with spikelets densely arranged along their length, closely pressed together.

Distribution & Habitat: A widely distributed and adaptable grass, common on roadside verges and fallow land, and in forests and the edges of woodlands. Ditch Millet can survive in waterlogged soils and seasonally damp areas, and is typically more common in wetter, more humid habitats within East Africa, although it can thrive in drought-prone areas. It grows from sea level to areas over 2,200m in altitude. The species is also widely grown for food in parts of Asia.

Ecology & Uses: An important forage and fodder species for livestock. It is heavily grazed at all stages of growth, and rapidly grows new leaves following grazing. The juicy, nutritious leaves are especially good for dairy cattle and milk production. Because it has a low fibre content, it is considered a more easily digested species.

The young grass has the racemes in a V-shape.

A highly variable inflorescence, but in East Africa it often has two racemes.

The soft, juicy leaves are very nutritious – a favourite of dairy cattle.

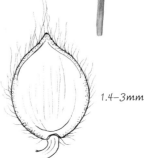

1.4–3mm

spikelet

Young leaves appear rapidly after rains.

The racemes spread further apart as the spikelets ripen.

This grass supports abundant grazing in areas where livestock are tethered or graze close to homesteads.

Kikuyu Grass

Pennisetum clandestinum   Height 3–45cm

Description: An attractive, robust perennial grass that grows in a dense mat-forming pattern, typically covering the ground in a lush carpet. The mats are made up of slender rhizomes, with creeping stems that readily root as they grow out and cover the ground. It can form trailing creepers up to a metre long. In wetter areas, it can grow into a dense, leafy sward up to 45cm tall. The older stems are covered with the remains of old leaves. Bright green leaves grow flat and folded and are soft and juicy to the touch, 1–15cm long. The inflorescences are hidden in the folds of the leaves at the tips of the stems, hence the scientific name *P. clandestinum* – meaning 'hidden' or 'secret'. Spikelets are long and narrow, 10–20mm long. The anthers are often extruded from the leaves on long, thin white stalks, dangling in the breeze as they shed their pollen.

Tiny seeds are hidden in spikelets, which are enclosed within the leaves.

10–20mm

seed

Distribution & Habitat: A common and widespread grass in the highlands, upland grasslands and moist zones of East Africa. It grows on well-drained, fertile soils and is often found on roadside verges, along streams in highland areas and on open land. Occurs in areas that are 1,300–3,300m above sea level.

Ecology & Uses: An important grass for both livestock and wildlife. Kikuyu is considered to be one of the most palatable, nutritious grasses. It is widely cultivated as a fodder grass and managed as a lawn grass in the highland regions of East Africa, and in many other parts of the tropics. This species can grow from cuttings and runners, making it easy to establish on bare ground. It grows rapidly, providing ground cover and preventing soil erosion.

The fine white 'threads' are filaments – stalks that bear the anthers.

Creeping stems take root easily as they spread along the surface.

Kikuyu Grass has the amazing ability to grow rapidly and cover ground, providing a lush and nutritious grazing and ground cover.

Pom-pom Wire Grass

Pennisetum mezianum   Height 30–130cm

Description: A dense, clump-forming perennial grass that grows in a tufted fashion from a short, woody, irregularly shaped rhizome. The leaves and stems sprout out untidily from the woody base. Leaves are either flat or folded, 2–15cm long. The inflorescence is a dense, spike-like panicle that sits on the end of a long stem, looking like a pale green pom-pom when young and ripening to a dark, dingy brown colour. These 'pom-poms' can persist on the plant for many months after rains.

Densely packed spikelets are held aloft on long, wiry stems above the plant.

Distribution & Habitat: A widely distributed grass in the mid-altitude grasslands of East Africa. It is most abundant on black-cotton clay soils, where it can form dense, continuous stands. Pom-pom Wire Grass also thrives in semi-arid regions and bushland, where it grows interspersed with other species. In certain parts of East Africa, this grass can become a tough, dominant species that gives a characteristically untidy appearance to the grassland.

Ecology & Uses: This species is well grazed by livestock and wildlife when it is young and sprouting new leaves. As the plant matures and becomes tough and woody, it is avoided by grazing animals as it is no longer palatable. In times of drought, the persistent leaves are an important source of food, making this species the last to be grazed once other grasses have been finished off completely or died back. Dung beetles of the genus *Neosisyphus* attach their dung balls to the stems, which is different from most dung beetles that bury their dung balls in the ground.

Young and ripening inflorescences often occur side by side.

3–4.5mm

spikelet

Some dung beetle species attach their dung balls to the stems of this grass.

Dense, slightly untidy clumps are typically found on wetter, clay soils, and are often associated with subterranean termite mounds.

Napier Grass

Pennisetum purpureum   Height 1–6m

Other name: Elephant Grass

Description: A tall, leafy robust grass that can grow up to 6m high, forming large, impenetrable clumps. The stems are thick and heavy, with the leaves clustered along the ends. Leaves are broad and bright green, 40–120cm. The inflorescence is an attractive spike-like panicle with the spikelets clustered along its length. It is green or purple when young, fading to brown or yellow as it ages and ripens. The individual spikelets are round in shape, surrounded by dense, coarse hairs.

Distribution & Habitat: A common and widespread grass in the wetter, higher-rainfall areas of East Africa. It is naturally found on deep, moist and fertile soils in riverine areas, along forest edges and in woodlands. Napier Grass does well from sea level to areas above 2,000m in altitude. It is also widely distributed across tropical Africa and is cultivated worldwide.

Ecology & Uses: As one of the most important fodder grasses, it is widely cultivated across East Africa and the world. This grass has a high leaf production and vigourously sprouts new leaves after harvesting or grazing. It is also palatable and easily digested, making it an excellent source of fodder for livestock. Napier Grass can be used for silage, and some varieties are useful for providing ground cover on bare ground and for protecting the edges of wetlands. The species is also cultivated ornamentally in gardens.

The spike-like inflorescence has spikelets with long, coarse awns.

Lush stands of Napier Grass are found on moist, fecund soils.

4.5–7mm

spikelet

Napier Grass sustains much of the small-scale livestock in East Africa; seen here in flower – a rare sight as it is typically cut back to allow for more leaf production.

Riparian Wire Grass

Pennisetum riparium   Height 10–150cm

Description: A tall, robust perennial grass that grows in dense, compact clumps with the stems closely packed together. Leaves are flat or slightly folded, narrow, dark grey-green, 4–30cm long. The inflorescence is an erect, straight spike that starts out pale green, becoming yellow or darker over time. The dark brown anthers are an interesting feature because they persist after flowering, creating a mottled appearance from a distance.

Distribution & Habitat: A common and widespread grass in East Africa. It typically grows in riverine areas, along forest edges and on roadside verges, and in wetter spots within grassland in the highlands. This wire grass is found at a wide range of altitudes, from 1,400–2,600m above sea level.

Ecology & Uses: An important grazing species that is utilised by both wildlife and livestock. Cattle graze on it when it is young, but they also seek out the mature plants. It is heavily grazed by game on the plains when other grasses have been depleted or have dried out. As a fairly tough species, it is an important ground cover and a source of stability in grassland.

Young (left) and more mature (right) – flowering stages can look quite different, but often both inflorescences can be spotted in a single clump.

The dark brown anthers persist on this plant when it flowers.

6–10mm

spikelet

This grass can remain in flower for many months once the rains have started, as it has a much longer growing period than most other grasses.

Roadside verges that accumulate water from runoff are one of the areas where this species thrives.

Kopje Wire Grass

Pennisetum squamulatum   Height 60–200cm

Description: A robust, stocky grass that grows in dense, almost impenetrable clumps. Leaves are flat, slightly stiff, and quite hard to the touch, up to 40cm long. The inflorescence is a straight, spike-like panicle, 10–25cm long. It is green or yellow-green when young, often with the anthers dangling and appearing to cover the surface when in flower. The inflorescence ages to a lovely golden-straw colour, and persists even once the seeds have been shed.

Inflorescences retain their bright colours for some time, even as the spikelets are shed when ripe.

Distribution & Habitat: A widespread, locally common grass in East Africa. It is found in abundance, growing in large leafy stands on rocky hillsides and isolated inselbergs, and in bushland habitats, at altitudes of 900–2,200m.

Ecology & Uses: An important grass for many different species of herbivore when other grasses are depleted or absent. As it grows in areas where its sturdy rhizomes and roots can reach water in rocky crevices, it remains leafy and green for longer periods of time. It is sought out by elephant, klipspringer and rock hyrax in the dry season. The persistent stems provide nest sites for insects, including solitary bees, wasps and other small cavity-nesting creatures.

The inflorescence is covered with a mass of anthers at the flowering stage.

3.5–7.5mm

spikelet

Many different insects feed and shelter in Kopje Wire Grass, including the fast-flying Roger's ranger (*Kedestes rogersi*).

Atop the kopjes – protected from large herbivores such as elephants – it forms large, densely leafy clumps.

Silver-plumed Wire Grass

Pennisetum stramineum   Height 40–130cm

Description: A tough, wiry perennial grass that grows from a short rhizome and branches to form large, dense clumps. Leaves are 2–20cm long and flat, tapering to a fine pointed tip; typically held upwards at an angle. They are fresh yellowy green when young. The inflorescence is an upright silvery spike (the 'plume'). Spikelets are covered in bristles that give the grass its distinctive appearance.

Distribution & Habitat: A fairly common and widespread grass in East Africa, typically found in abundance on black-cotton clay soils, but also in rocky areas on red soils. It grows at altitudes of 1,000–2,500m.

Ecology & Uses: Most often found growing in association with the subterranean mounds made by termites (*Odontotermes* spp.). When viewed from the air, these mounds form distinctive ghostly circles in the grassland – something like a watermark that is repeated at regular intervals. Both livestock and wildlife graze this species when it is young, but avoid it as it ages and becomes very tough. Silver-plumed Wire Grass is used by birds for weaving their nests, and many different insect species make homes in the older clumps.

Bristles around the spikelet become more silvery as the inflorescence ripens.

The fine, narrow spikes appear later in the rainy season, once the grass is well established.

4–5.5mm

spikelet

The leaves of this tough, wiry grass are grazed when young by many species, including the Grevy's zebra.

Young leaves of this grass are brighter green, as seen in the foreground.

Montane Pentameris

Pentameris minor
(= *Pentaschistis minor*)   Height 10–30cm

Description: An elegant, perennial grass that grows in loose tufts. Leaves are 2–7cm long, narrow, folded and fairly stiff, often slightly clumped or held together at the base of the plant. The inflorescence consists of a somewhat narrow, slightly open panicle with the spikelets held on branches. The spikelets are an attractive, eye-catching silvery colour, held on a stem above the leaves. Short awns are attached to the base of each spikelet.

Distribution & Habitat: A fairly common and widespread grass in the montane and high-altitude regions of East Africa, including the Aberdare Mountain Range and Mounts Kenya, Elgon and Kilimanjaro. Montane Pentameris has several closely related species that are typical of high-altitude and moorland habitats in East Africa. The *Pentameris* grasses are some of the most hardy, common plants in areas 3,000–4,000m above sea level. This species is also found growing among loose rocks and scree at close to 5,000m, making it one of the highest-altitude plants in the region.

Ecology & Uses: An important grass of the montane and moorland zones, it often grows among the tussock grasses (*Festuca*). It is locally grazed seasonally by common duiker and rock hyrax. As it can survive in extremely cold and Afro-alpine zones, it is an important species for providing ground cover and stabilising the fine soils in these high-altitude areas.

Spikelets are tinged pinkish purple when young, and turn silvery grey as they ripen.

3.7–4.7mm

spikelet

A stand of young Montane Pentameris turns more silvery as the spikelets ripen.

This grass is often found growing at the very highest altitudes, among mosses and seemingly on bare rock, where it survives freezing conditions.

Bottlebrush Grass

Perotis patens   Height 30–60cm

Other name: Catstail Grass

Description: An attractive and distinctive annual grass with a loosely tufted growth form and wiry stems. In drier areas plants tend to be smaller overall. Leaves, 1–7cm long, are a narrow spear or oval shape with a tapering tip. The inflorescence is a striking, easily recognisable dense spike with the spikelets neatly clustered along the stem. Spikelets are rounded, with a rough appearance. The fine, reddish-purple awns, 9–17mm long, are symmetrically arranged in a tidy fashion, held out at right angles from the main stem.

Neatly arranged spikelets look like the teeth of a fine comb; they spread out more as they ripen.

Distribution & Habitat: A widely distributed grass found in a range of soils in the drylands, in grassland and bushland, the arid and semi-arid regions, and also on rocky slopes. It grows and thrives in areas from sea level up to 1,650m.

Ecology & Uses: Not heavily grazed by livestock or wildlife, though it is grazed in times of drought or when other grasses are scarce. As it grows well and establishes rapidly on recently grazed, burnt or bare ground, this grass is an important pioneer species that provides ground cover and prevents soil erosion.

This grass grows rapidly on bare ground, helping to provide important soil cover.

1.2–3mm

spikelet

Bottlebrush Grass can form attractive, small clumps following good rains in dryland areas.

Reed Canary Grass

Phalaris arundinacea   Height 50–150cm

Description: An elegant, tufted perennial grass that grows from a hidden creeping rhizome. It occurs singly or in tall clumps depending on its location and the availability of moisture in the soil. Leaves are bright green, 7–40cm long. The inflorescence is a distinctive, attractive and often slightly drooping panicle, typically green with red and purple markings. It appears to be bright red in younger plants when viewed from a distance, turning straw coloured as it ripens. The individual spikelets are densely packed together, 3.5–7mm long.

The dense, dangling, purple inflorescences may appear bright red from a distance.

Distribution & Habitat: A fairly common and widespread grass of the highland areas in East Africa. It grows along streams, around the edges of swamps, bogs and wetlands, and in damp spots within the highland-grassland and moorland vegetation zones.

Ecology & Uses: This leafy perennial highland species is grazed by livestock and wildlife, often in the dry season when other grasses have been consumed or died back. It does well in waterlogged areas, but also has a degree of drought tolerance and can grow on poor soils. A number of varieties and cultivars have been developed as ornamentals, as well as for land restoration and fodder production.

3.5–7.5mm

spikelet

One of the most resilient highland grasses, this species can be found lining many highland roads and streams.

The inflorescences keep their colour even as the leaves start to age. Eventually they turn pale and disintegrate.

Common Reed Grass

Phragmites australis   Height 1.5–3.5m

Description: A tall, striking and distinctive perennial species that grows in large, open to dense stands. The stems are long and wiry, with the leaves loosely arranged along their length. Leaves, 20–60cm long, are pale green with fine, tapered, pointed tips. The inflorescence is a distinctive feathery panicle with spikelets clustered together, readily visible and recognisable from a distance. The spikelets, 12–18mm long, turn from pale pinkish purple to straw coloured as they ripen.

Distribution & Habitat: A common and widespread grass in East Africa and prevalent across the tropics. This aquatic plant grows in fresh water and in partially saline conditions; found on the edges of lakes, slow-flowing streams and rivers, and in wetlands. It is often established around dams and in constructed wetlands. A number of closely related species occur in the coastal zones and deltas. Common Reed Grass is found mainly at 500–2,000m above sea level.

Ecology & Uses: A very important aquatic species that provides structure and habitat wherever it grows. Many birds, insects and other species find shelter, nesting sites and refuge within stands of this grass. It helps to provide stability in wetlands, moderates the effects of runoff and flooding, traps silt and reduces erosion. In addition, it acts as a natural filter that purifies water. The stems are used to make local handicrafts such as mats, lampshades, baskets and other household wares.

Just a single branch of the large, dense inflorescence can be composed of thousands of spikelets.

12–18mm

spikelet

This important grass helps to stabilise riverbanks in East Africa.

The inflorescence has many tiny closely packed spikelets along the stems.

When young, the leaves have sharp points that can give a painful prick.

Large stands of this grass are home to many different birds and other animals.

Herringbone Grass

Pogonarthria squarrosa   Height 20–150cm

Description: A tufted perennial grass that grows in dense clumps and is highly variable in size; in most of East Africa it is a fairly short, straggly grass. Leaves are flat and narrow, sometimes folded slightly inward, 4–30cm long. The inflorescence is a stiff, ascending shape with the spikelets arranged on short, straight branches, leaning away from the central axis in a symmetrical pattern, and tapering as they reach the tip.

Distribution & Habitat: A widespread and common grass all over East Africa, found across many different habitat types at a range of altitudes from 500–2,000m. It is also found in other parts of the continent, including the Horn of Africa and South Africa.

Ecology & Uses: This grass is not typically grazed by livestock. It often establishes on the edges of tea fields and other cultivated areas where it is important in controlling runoff and reducing soil erosion. As it grows in a range of soil conditions, including in sandy soil, it is an important pioneer species and helps to stabilise bare ground, paving the way for other plants and grasses to establish themselves.

The compact inflorescence has a symmetrical appearance and is easily recognised when the grass is mature.

3.3–7.5mm

spikelet

Young and mature inflorescences seen side by side

Herringbone Grass is one of the smaller species that help to provide ground cover and protect soils.

Forest Shade Grass

Pseudochinolaena polystachya   Height 10–30cm

Description: A delicate, pretty, creeping annual grass that grows in a trailing fashion with creepers up to 1m long, rooting intermittently at the nodes, often with some roots exposed along the length of the plant. Leaves are typically held flat and can form loose mats. Shoots bearing the inflorescence rise above the leaves on short stems. The inflorescence consists of loosely arranged spikelets that are closely pressed along the stem. Spikelets bear bristles and are loosely attached, breaking free and falling off with ease.

Distribution & Habitat: A common, widespread grass in forests and woodlands across most of the wet tropics. It is often found in wetter forests in western Kenya, western Tanzania and Uganda.

Ecology & Uses: Not typically grazed by livestock. It can be locally abundant within forest habitats, where it is utilised by birds, rodents and forest antelopes, including duiker. Forest Shade Grass is a favoured perch for insects sunning themselves. As it grows along forest and woodland paths, even in deep shade, it is important for controlling erosion and slowing the runoff from rain.

Spikelets are loosely arranged but closely attached to the stems.

The anthers and stigmas are visible on the spikelets during flowering.

The glumes bear bristles and readily detach as the grass ripens.

3.5–5.5mm

spikelet

Perhaps one of the softest carpets in forest shade is made by this grass.

Sehima Needlegrass

Sehima nervosum   Height 30–150cm

Description: An elegant, medium-sized perennial
grass that forms narrow, thin clumps. Leaves are
sparsely distributed, 5–40cm long. The inflorescence
consists of a narrow spike that is typically held upright
in younger plants, before arching gracefully at an angle
as the grass matures. Spikelets on the young spikes
are arranged in a graceful pattern, with regular purple
markings along their length. Older, mature spikelets
have a more untidy appearance.

*This grass has
beautifully patterned
glumes when young,
with the spikelets
and awns neatly
pressed together.*

Distribution & Habitat: A very widespread grass
species that occurs in most of East and southern Africa
and the Greater Horn of Africa. It is a classic species of
arid and semi-arid regions. Sehima Needlegrass typically
grows interspersed with other grasses, and thrives on
stony or rocky soils. The species is also found in the
Middle East, Asia and Australia.

Ecology & Uses: This grass is not heavily grazed by
livestock because it has a relatively low leaf production.
However, it is seasonally grazed in arid areas by wildlife,
including oryx, Grevy's zebra and gazelles. It establishes
easily, even on disturbed land in arid areas, and so is an
important pioneer species that creates the necessary
structure and opportunity for succession by more
palatable grass species.

Sehima Needlegrass in flower

As it ripens, the grass becomes twisted and the awns
more exposed.

6–10mm

spikelet

Stands of this grass typically grow together with Buffel Grass and Kenya Three-awn Grass, both of which can be seen here; Sehima Needlegrass often stands taller than these other species.

Broad-leafed Bristle Grass

Setaria megaphylla   Height 1–3m

Description: A tall, robust perennial grass that forms large, dense, leafy clumps and grows up to 3m tall. Leaves, 15–80cm long, are broad, 1–2cm wide, with distinctive 'pleats' along their length. The inflorescence is an elegant panicle, held above the leaves or arching at a graceful angle. Spikelets, 2–3.5mm long, are arranged on short branches; those on plants in shade tend to be greener while in sun they are more purple.

Distribution & Habitat: A common and widespread species of grass that grows across East Africa and is also widespread in other tropical regions of the world. It is typically found in slightly wetter areas within forests and woodlands, and along streams, paths and roadside verges.

Ecology & Uses: As a very leafy grass in forest habitats, it is grazed by a variety of animals, including buffalo, elephant, duiker, bongo and other forest-dwelling antelopes. The leaves and stems are sometimes used as thatching material for roofs or for temporary shelters in the forest.

The large, elegant inflorescence often persists, even as the spikelets mature and drop off the plant.

2.2–3.5mm

spikelet

Stalk-eyed flies often sit on the leaves to sun themselves and watch for mates.

The undersides of damaged leaves often reveal the caterpillars that have been feeding on them, like this larva of an evening brown (*Melanitis* sp.).

Adult evening brown butterflies can often be found sheltering or laying eggs on this grass.

When the plant is in flower, the clusters of yellow pollen-bearing anthers are visible.

This grass can grow among the roots of large forest trees where it provides shelter for many different creatures.

Yellow Bristle Grass

Setaria pumila   Height 5–130cm

Description: A straggling or loosely tufted annual grass that grows in somewhat threadbare clumps. It varies greatly in height depending on where it grows, tending to be shorter and more compact in drier sites. Leaves are narrow, 3–30cm long. The inflorescence is a straight, symmetrical spike with the individual spikelets packed closely together along its length. Spikelets are green when young, fading to pale brown. The most striking feature of this species is the bright yellow or yellow-brown bristles that persist on the inflorescence; these are finely arranged, held at an angle to the main stem.

Distribution & Habitat: A very widespread and common grass throughout East Africa. It grows and thrives on a wide range of soils and conditions, and does especially well on recently disturbed ground, in roadside verges and ditches, and as a weed in fallow land.

Ecology & Uses: This widespread grass in diverse grasslands is grazed by both wildlife and livestock. It is also important grazing for cattle and sheep in mixed-farming systems in the mid-altitude regions of East Africa. It is drought tolerant and, as a pioneer species, it helps to stabilise bare ground and prevent erosion.

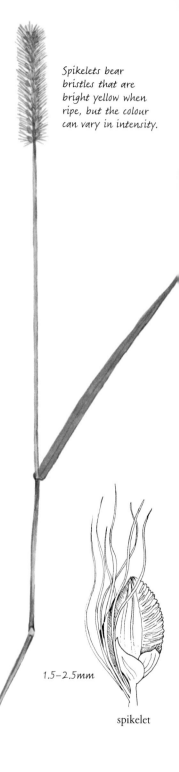

Spikelets bear bristles that are bright yellow when ripe, but the colour can vary in intensity.

1.5–2.5mm

spikelet

The neat, compact spikes make this grass easy to recognise.

Yellow Bristle Grass is one of the most important pioneer species on bare or recently disturbed ground, including in old quarries and mining areas.

Golden Bristle Grass

Setaria sphacelata   Height 1–2m

Other name: Golden Millet

Description: A tall, elegant perennial grass. Highly variable, there are several subspecies that occur across East Africa. The plant forms dense stands and dominates grassland, growing in clumps from a creeping rhizome. Leaves are 10–50cm long, flat and abundantly produced. The inflorescence is a single pretty spike that is often golden brown in colour, but varies from light green to pale brown, with the spikelets densely packed along its length.

Distribution & Habitat: One of the most widespread and common grasses in East Africa. It grows and thrives in a wide range of areas in open, wooded and highland grassland, in bushland, and on black-cotton clay soils. Golden Bristle Grass often forms dense, continuous stands that dominate the grassland. It is found from sea level up to around 3,000m.

Ecology & Uses: An important grazing resource for both livestock and wildlife. The dense leafy growth is attractive to both traditional grazers (cattle and zebra) and mixed feeders (buffalo). The seeds are eaten by a wide range of birds, and several insects use this grass as a host plant during their larval stage.

The bright, golden bristles and long, elegant spikes make this one of the most familiar grasses of East Africa.

1.5–3.5mm

spikelet

Many different insects, including blister beetles, visit the young inflorescences.

Young and older ripening spikes of this grass often appear together on the plant.

A variety of animals, including these African buffalo, love to graze on this leafy, nutritious grass.

Bur-bristle Grass

Setaria verticillata   Height 10–100cm

Other names: Rough Bristlegrass, Bristly Foxtail Grass

Description: A medium-sized annual grass that grows in an untidy trailing fashion, often partly draped over or woven among other grass and herbaceous species. Leaves, 5–35cm long, are straight and can be smooth or slightly hairy. The inflorescence has distinctive densely packed individual spikelets with bristles that readily cling to clothes, fur, skin and even other plants.

Distribution & Habitat: A common and widespread grass species in East Africa and most of the tropical and subtropical areas of the world. It grows on a range of soil types and at different altitudes, from near sea level moving up into highland areas 2,200m in altitude. Bur-bristle Grass is often found in recently cultivated areas and fallow land; it is regarded as a weed in actively cultivated areas.

Ecology & Uses: A grass of moderate grazing value for livestock in most areas. It tends to favour roadside verges and recently disturbed ground, which makes it an important pioneer species. The extremely clingy seeds are widely and often unwittingly dispersed, including by humans. In many places children will use the inflorescences in games, sneakily attaching them to the clothing of unsuspecting recipients.

The seeds of this grass have been used as a source of food in some areas during times of famine.

The broad, flat leaves are eaten by herbivores when young.

The bristles are covered in short, tenacious hooks that give this grass the ability to cling to almost anything that brushes up against it.

3–8mm

spikelet

This leafy grass is often hidden among other plants when young.

When young, the bristles are purplish and spikelets bright green.

In many places this grass establishes on bare ground, helping to create cover and protect the soil.

Wild Sorghum

Sorghum versicolor   Height 25–250cm

Description: An elegant, graceful annual grass that grows in an upright, erect form, typically 1–2m tall in most parts of East Africa. Leaves are flat, bright yellow-green in colour, 10–30cm long. The inflorescence is a striking, open panicle that has a broadly oval, tapered shape, 5–25cm long. The spikelets are borne on stems that grow in whorls from the central main stem. Individual spikelets are greenish and covered with reddish-brown or brown hairs, 25–40mm long.

Distribution & Habitat: A common and widespread grass in East Africa. It is typically found in slightly wetter areas on moist soil, in bushland, woodland and tall grassland, at the edge of fallow and cultivated areas, and in riverine habitats and seasonal waterlogged pans. Found from near sea level to areas of 2,000m in altitude.

Ecology & Uses: An important grass that is utilised by both livestock and wildlife, especially when it is young. Large herbivores, including elephant, will graze on it where it is available. As it tends to grow in wetter areas, it often stays fresh and green when other grasses are drying up. The seeds are a favourite of many birds such as weavers and bishops. Wild Sorghum has been used as a trap crop for the parasitic *Striga* plant (witchweed) and also for stemborers. This strategy of using certain plants to draw pests and weeds away from a main crop is part of the push-pull approach of integrated pest management, commonly used in East Africa to boost the cultivation of maize and sorghum.

Cultivated sorghum, *Sorghum bicolor*, is an important crop in dryland areas of East Africa.

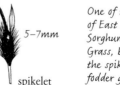

5–7mm

spikelet

One of the most recognisable of East Africa's grasses, Wild Sorghum is similar to Tall Guinea Grass, but can be told apart by the spikelets. It is an important fodder grass in some areas.

The spikelets turn black as they ripen and the awns help to disperse the seeds.

When in flower, the spikelets are more closely clustered together and have a more compact appearance.

Tiny Dropseed Grass

Sporobolus festivus   Height 20–30cm

Description: A short, sparse perennial grass that grows in thin, wiry clumps, sometimes even reaching up to 60cm in height. Leaves are flat and short, 2–7cm long. The inflorescence consists of a delicate symmetrical panicle with tiny spikelets spread out all over it. The stems supporting the spikelets are arranged in whorls, and are bright reddish in colour. Individual spikelets are purple-red when young and fade to grey-green as the plant matures.

Distribution & Habitat: A widespread and common species across East Africa and most of the African tropics. It is found in woodland and bushland, and on rocky pans and in crevices. This grass typically grows on shallow soils and in association with Hairy Love Grass.

Ecology & Uses: This fairly small and light grass is not widely grazed by livestock, but is consumed by wildlife, including warthog, bushbuck and dik-dik. Where it grows on rocks and shallow soils, it is important as a soil stabiliser and it helps to prevent erosion.

This delicate grass has the potential to be used more in gardens and natural borders for landscaping.

The inflorescences are extremely fine and delicate, with tiny spikelets.

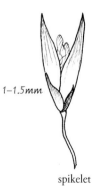

1–1.5mm

spikelet

Though tiny, this grass is a tough species that can survive even in some of the most exposed sites, like here on a rocky outcrop.

Bushveld Dropseed Grass

Sporobolus fimbriatus   Height 25–100cm

Description: A tough, wiry perennial grass that grows in loose, open tufts from a short underground rhizome. Leaves, 10–30cm long, are narrow and slightly rough to the touch on the underside. The inflorescence consists of a pretty, somewhat tapered, open panicle, with many branches in loose whorls along the main stem and tiny spikelets arranged at the tips of the branches. The spikelets are a dull grey-green in colour, 1–2.2mm long. The lower branches of the inflorescence often emerge somewhat clustered together from the base of a leaf.

The inflorescence is very fine and feathery, with the tiny spikelets packed close to the branches.

Distribution & Habitat: A common and widespread grass in East Africa, it is also widely distributed through tropical and southern Africa. It grows in fallow areas and recently bared ground, also underneath trees and shaded bushes, and among rocks. Bushveld Dropseed Grass does well on slightly more moist soils from the coast all the way up to areas over 2,000m in altitude.

Ecology & Uses: A useful grazing species for both livestock and wildlife. Where it grows in more hidden places or among rocks, it is often grazed when other grasses have dried up or been grazed down. The leaves are very nutritious and often sought out by animals, including bushbuck and other antelopes. The seeds are consumed by seed-eating birds such as purple grenadiers, cordon-bleu finches and species of waxbill.

Livestock graze on this grass when it is young and leafy.

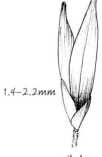

1.4–2.2mm

spikelet

In more sheltered areas the grass can form large, leafy clumps with the tall panicles rising above the leaves.

Catstail Dropseed Grass

Sporobolus pyramidalis   Height 40–160cm

Other name: Giant Ratstail Grass

Description: An elegant and upright medium to tall perennial grass. The narrow leaves can reach up to 50cm long, and are flat or slightly curved, with a fine, pointed tip. The inflorescence is a striking, symmetrical panicle that has a tapered, pyramid-like shape. The spikelets are tiny, packed densely along the whorled spikes, yellow-green when young and quickly turning dark grey-green.

Distribution & Habitat: A very widely distributed and common grass species that grows on a range of soil types and at different altitudes. Growth patterns are variable; it can grow in individual clumps mixed with other species, and also in dense stands by itself. It often grows in waterlogged spots or at the edges of seasonal rocky pools and streams, and is also found in recently disturbed, cultivated or fallow places. It grows in areas from sea level up to over 2,500m in the highlands.

Ecology & Uses: A tough and wiry species that is not typically grazed by cattle in wetter highland areas, but it is an important grazing species for sheep and goats in drylands. The growth pattern and tough nature of the mature clumps make this an important refuge for insects and spiders. Warthogs and bushpigs will snuffle around to feed on the roots, and the seeds are consumed by many different birds and rodents.

Many different herbivores graze on this grass at all stages of its growth.

A mature inflorescence has a tapering open panicle.

1.7–2mm

spikelet

When young, the panicle is compact and slightly curved. This species is spreading in some areas outside of its native range in Africa and the Middle East.

This grass can thrive in rocky stream beds as well as more open areas.

Coastal Turf Grass
Stenotaphrum dimidiatum   Height 10–30cm

Description: A creeping perennial grass that forms dense mats covering the ground. It tends to grow taller and more loosely in partly shaded spots, and more tightly in open areas with full sunshine. Leaves, 5–20cm long, are flat with rounded tips. The inflorescence is distinctive, consisting of a spike that is held upright or at an angle, and is somewhat thickened and fleshy, feeling slightly corky to the touch. The small, oval-shaped spikelets are arranged along the length of the spike.

Distribution & Habitat: Occurs primarily in the coastal regions of Kenya and Tanzania in East Africa. It is found in coastal woodlands and glades, open areas, and close to the seashore in areas behind the dunes. This species thrives on sandy soils and can tolerate partly saline conditions. It grows in areas from 0–100m in altitude.

Ecology & Uses: A nutritious and leafy grass that is widely grazed by both livestock and wildlife. It can form a dense pasture, and may be managed as a grazing and fodder species for cattle. Coastal Turf Grass offers great potential for better livestock management in zero-grazing or intensive livestock farming. In many coastal regions, this species, together with the related but introduced species *Stenotaphrum secundatum*, is often used to establish lawns around houses, hotels and resorts.

The flattened inflorescence has spikelets arranged along its length.

4–5mm

spikelet

In times of heavy rainfall, the spikelets are often infested by a fungus and become dark and powdery.

The rounded leaf tips are diagnostic.

If not heavily grazed or mown, this grass can form a lush stand, which can be used as fodder for livestock.

Plume Grass

Tetrapogon roxburghiana
(= *Chloris roxburghiana*)
Height 50–150cm

Other name: Feather-duster Grass

Description: This striking perennial grass is one of the most recognisable and beautiful of the East African grasses. It grows in dense stands with leaves tightly pressed together and overlapping. Leaves are 10–45cm long. The distinctive inflorescence is bright purple to pink, fading to pale pink and light green, and eventually to pale and creamy straw coloured. The spikelets are densely clustered on the inflorescence and detach in numbers when the grass is maturing; they have 3 or 4 wispy awns of equal length.

Distribution & Habitat: A widely distributed species across East and southern Africa, the Greater Horn of Africa and parts of India. It is found in bushland and woodland, and thrives both in glades in these habitats as well as in more open areas, including on disturbed ground.

Ecology & Uses: An important species for grazing when young by livestock and wildlife in dryland areas. This is a robust and fast-growing grass that regenerates quickly, and because it is highly visible, it is often targeted for grazing by herders. It is also used as an ornamental plant, both in gardens and flower arrangements.

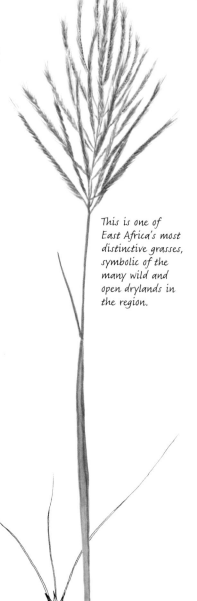

This is one of East Africa's most distinctive grasses, symbolic of the many wild and open drylands in the region.

2–3mm

spikelet

Younger inflorescences are a darker shade of purple, fading to a bright pinkish colour.

Plume Grass adds a festive air to the landscape following good rains.

Tender Finger Grass

Tetrapogon tenellus   Height 20–80cm

Description: An attractive, variable grass that typically grows as an annual, and occasionally establishes itself as a perennial. The growth form is erect, with stems that can reach up to 80cm. Leaves, 10–24cm long, are narrow with a fine, tapering point. The inflorescence is distinctive, with typically one and sometimes two attractive spikes. These have a graceful, symmetrical outline with the individual spikelets arranged in a neat, alternating fashion along the length of the spike. Spikelets are green with purplish-red awns when young, fading to a pale, creamy colour.

Distribution & Habitat: A common and widespread grass in the drylands, woodlands, and arid and semi-arid regions of East Africa. It often grows sparsely among other grasses and can grow in the partial shade of bushes and small trees. This species is also found in other dryland and tropical regions, and across the Greater Horn of Africa and the Middle East into India.

Ecology & Uses: As one of the grasses that thrive in drylands and arid and semi-arid areas, this grass is an important and nutritious source of grazing for livestock and wildlife. The attractive, easily recognised spikes signal good grazing to herders moving livestock around. Various birds and insects feed on the spikelets in all growth stages, from the young green spikelets to the mature dried spikes.

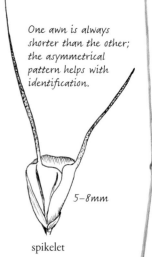

The inflorescence is idiosyncratic, with mostly one but sometimes two spikes.

One awn is always shorter than the other; the asymmetrical pattern helps with identification.

5–8mm

spikelet

This grass can survive in some of the hottest, most exposed sites, where it is an important pioneer species.

The neat, alternating arrangement of the spikelets makes this a very easy grass to identify.

Red Oat Grass

Themeda triandra   Height 30–300cm

Description: One of the most recognisable and striking of all East Africa's grasses, this grass can vary greatly in height. The plant forms dense, leafy clumps. Leaves, up to 30cm long, are flat and bright lemon-green when young, darkening slightly as the plant ages. The inflorescence consists of a number of elegant drooping clusters of spikelets. The spikelets have a characteristic insect-like or fishing-fly appearance. They are held at a drooping angle, often waving gently in the breeze. Young plants are bright green, tinged with red. As the grassland matures, this species turns a dazzling coppery red colour that is distinctive from a distance.

Distribution & Habitat: A very widely distributed grass species found throughout East Africa on a variety of soils. It often grows in dense stands forming attractive swathes across the plains. This species is typical of all major grassland and savanna regions, including the Mara–Serengeti ecosystem, Laikipia Plateau, the central highland and areas adjacent to the Rift Valley. It can grow and thrive at a range of altitudes, from coastal glades near sea level to highland grassland at over 3,000m.

Ecology & Uses: One of the most palatable grasses, and a favourite grazing for both livestock and wildlife wherever it occurs. The leafiness, robust growth and nutritious composition make it one of the most economically important wild grasses of the region. The spikelets bear awns that enable the seeds to implant themselves in the soil, where they can elude seed-eaters and survive droughts and fires, remaining viable for a long time. The buried seeds sprout after rains and grow rapidly to form new plants.

Long awns help to bury the seeds in the soil, where they can survive droughts and fires.

6–15mm

spikelet

Young inflorescences look like clusters of perched winged insects about to take flight.

As this grass ripens, the spikelets spread out and form a more untidy pattern of growth.

Fields of Red Oat Grass are a favourite grazing for many herbivores, including elephant.

Carrot-seed Grass

Tragus berteronianus Height 10–20cm

Description: A short, compact annual grass that occasionally grows up to 40cm in height. Leaves, only 1–6cm long, are short and fairly stiff, with a wavy margin fringed with hairs. The inflorescence is very distinctive and forms a narrow, straight spike with closely clustered spikelets arranged along its length. The spikelets are prickly, covered with recurved spikes that readily attach to a range of surfaces.

Distribution & Habitat: A common and widespread grass of the drylands, rangelands, and arid and semi-arid regions of East Africa, northern Africa and the Middle East. This grass is typically found on sandy or stony soils, at the edges of road verges, and on trampled or disturbed ground. It grows at a range of altitudes from sea level up to 2,000m.

Ecology & Uses: Grazed in drylands when young by rodents, ground squirrels and dik-dik. This important pioneer species establishes itself on bare ground where it forms a very effective ground cover. It clings tenaciously to clothing, socks, soles of shoes, fur and feet, which enables it to be widely dispersed.

Prickly spikelets are paired with a smaller, hidden spikelet that is attached to the larger one.

2–3mm

spikelet

Once the prickly spikelets of this grass take hold, it takes quite an effort to detach them!

Spikes rise up from leafy sheaths after rains.

Carrot-seed Grass is one of the most resilient and tough grasses in the region; it can establish easily on bare ground, which provides stability and cover – conditions that allow other species to start growing there.

Garden Urochloa

Urochloa panicoides Length 10–100cm

Other names: Liverseed Grass, Garden Signal Grass

Description: A stocky, creeping annual grass that typically grows in a flattened position, pressed along the surface of the ground. Leaves, 2–25cm long, have slightly untidy, wavy margins, with hairs at the base. The inflorescence consists of a variable number of finger-like spikes, held at angles from a narrow stem that arises from the leaves. The spikelets have a somewhat flat, oval shape and are pale green when young, turning straw coloured as they ripen.

Distribution & Habitat: A very widespread, common grass in East Africa, also widely distributed across the tropics. It is found in many different kinds of soils, and can establish itself on waterlogged clays and dry, rocky ground. Found from sea level to 2,000m.

Ecology & Uses: An important source of grazing for both livestock and wildlife. As it can grow in a weed-like fashion in cultivated land, fallow areas and roadside verges, it is often grazed by tethered livestock. This fast-growing, easily established grass is important for stabilising bare ground and preventing soil erosion.

Racemes vary in number and are longer and more abundant in wetter conditions.

2.5–5.5mm

spikelet

On bare ground, following good rains, this grass forms a dense, leafy mat.

A robust and highly versatile grass, this pioneer species can grow and thrive in a wide range of conditions.

African Mountain Bamboo

Yushania alpina — Height 2–20m

Description: One of the most recognisable grasses of East Africa, this tall, elegant montane bamboo has graceful, hollow stems that rise from an untidy, woody clumped rhizome at the base of the plant. The stems are variable in height, mostly 5–10m tall. They are covered with sheaths when young, which have prickly bristles. Leaves, 5–20cm long, are narrow and tapered, borne on many straggling, drooping branches. The inflorescence consists of densely packed spikelets that are green with brown edges. The plants flower in synchrony following intervals of many years, sometimes decades.

Distribution & Habitat: Typically found in a distinctive bamboo zone on the mountains and highlands of East Africa, 2,400–3,000m in altitude. It also grows in forest patches and along streams at high altitudes throughout the region's montane habitats.

Ecology & Uses: An important part of the water tower ecosystems on mountains where it grows. Many different mammals, birds and insects use the bamboo for food, shelter or nesting. The endangered mountain gorillas in Uganda and Rwanda inhabit bamboo forests and feed on young shoots. This grass is widely harvested for use in local construction, fencing, enclosures for livestock and handicrafts by different communities.

This bamboo inflorescence grows hanging downward.

The tender shoots are an important part of the mountain gorilla's diet.

Ital Shanni

15–40mm

spikelet

This bamboo species flowers intermittently, sometimes with intervals of up to 20 years. Flowering happens synchronously across large areas; many insects appear at the same time, like this well-camouflaged seed bug.

Bamboo leaves in dappled sunlight are a classic feature of East Africa's mountains.

GLOSSARY OF TERMS

adventitious root root that grows from any part of a plant other than the main root

annual plant that completes its life cycle within one season, from germination to flowering and seeding

awn structure that is part of the grass spikelet; it resembles a hair or bristle

black-cotton clay soil soil type that is seasonally waterlogged and forms deep cracks in the dry season, found in many parts of East Africa

branch part on which the spikelets grow on the panicles; some grasses have varying branches off the main stem bearing the inflorescence

browse action of feeding by herbivores

browser animal that feeds primarily on trees and shrubs (see **grazer**)

C3 grasses grasses with the typical form of photosynthesis; 3 refers to the number of carbon atoms produced in the molecule from the first stage of photosynthesis

C4 grasses primarily tropical grasses that fix carbon in an intermediate compound with 4-carbon molecules; C4 grasses are more adapted to hot conditions

coral rock rocks derived from ancient coral reefs that are now above water and covered with vegetation

cover crop crop grown for the protection and enrichment of the soil

culm grass stem, typically round, partially hollow and with solid nodes (see **stem**)

endemic species that is found only within a limited biogeographic region

fallow, fallow land land that is inactive and uncultivated

fodder food, especially dried grass (hay), for livestock

glume bract at the base of a spikelet below the lemma and palea, typically the lowest part of the spikelet

grazer animal that feeds mainly on grasses (see **browser**)

Greater Horn of Africa region comprising Djibouti, Eritrea, Ethiopia, Somalia, Kenya, Tanzania, Burundi, Rwanda, Uganda, South Sudan and Sudan

hemipteran describing an insect of the order Hemiptera, such as an aphid, cicada or leafhopper

hindgut fermenters herbivorous animals that feed by packing their guts full of bulk vegetation and slowly digesting this in the hindgut with the assistance of microbes; zebra and elephant are hindgut fermenters

honeydew sugary secretion produced by hemipteran insects, offered as a reward to the ants that tend and protect them

Horn of Africa region comprising Djibouti, Eritrea, Ethiopia and Somalia

inflorescence complete flowering part, including the spikelets, of the grass plant

Intertropical Convergence Zone (ITCZ) band of weather activity associated with the equatorial regions of the planet

land reclamation process of restoring land that has been degraded through human activities or by overgrazing and trampling from livestock

'long rains' general term for the first rainy season in East Africa, typically starting in March and running through May

mixed-farming system system where crops are grown and livestock are kept

mixed feeders animals with characteristics of grazers and browsers; they graze and browse depending on time of year and availability of grasses; most mixed feeders prefer grasses when they are available

monodominant ecological pattern where a single species makes up the bulk of the individuals present in a population or given area

monsoon seasonal rainfall that comes to East Africa from the Indian Ocean, borne on the southeasterly trade winds, also known as the southeasterly monsoons

moribund in grasses, refers to old, standing growth that is low in nutrition and grazing value

mycorrhizal fungi specialised fungi that grow on the roots of plants in a symbiotic relationship

panicle form of inflorescence more spread out than the raceme or spike, with a number of branches from the main central stem pointing in different directions

perennial plant that grows, matures, flowers and seeds through multiple seasons; in East Africa, perennial grasses are able to survive through the dry season, sprouting new leaves and shoots when conditions improve

primary productivity production of calories by green plants through photosynthesis; this is the basis of the food web in grasslands and in all terrestrial habitats on Earth

raceme form of inflorescence with spikelets attached in even rows along the stem

rangeland open country used for grazing or hunting animals, typically with a seasonal or semi-arid climate

recurved curved backwards in a sickle shape

red lateritic soil soil type common in East Africa, volcanic in origin and one of the more fertile soils that support grassland habitats

rhizome horizontal underground stem that produces shoots along its length

ruminants herbivorous mammals that have specialised chambered stomachs to help digest plant matter; ruminants 'chew the cud' by regurgitating and re-chewing partly digested food

Sahel region comprising Burkina Faso, Cameroon, Chad, The Gambia, Guinea, Mauritania, Mali, Niger, Nigeria and Senegal

shifting cultivation type of subsistence farming naturally suitable for harsher environments, alternating crop and fallow through the seasons

silage form of fodder, where plants are maintained in a green, moist state

slash-and-burn method of agriculture in which existing vegetation is cut down and burnt off before new seeds are sown, typically used as a method for clearing forest land for farming

spike form of inflorescence where the spikelets are clustered along the stem

spikelet units of a grass that make up the inflorescence and contain the flowering parts, as well the seeds following fertilisation

stem in grasses, the central axis that supports the part of the inflorescence bearing the spikelets; typically round, partially hollow and with solid nodes (see **culm**)

stolon horizontal stem that grows in a creeping fashion along the surface of the soil and produces shoots along its length

sward expanse of grass covering an area

tussock form of growth that produces thick, dense, rounded clumps of grass

vlei shallow, natural body of water (South African English)

watershed drainage area that channels rainfall into streams, rivers, wetlands and lakes

whorl set of leaves, flowers or branches arising from the stem at the same level and encircling it

xeric adapted to survive and grow in dry or arid conditions

vector typically an insect or other arthropod that serves as a 'vehicle' to move a pathogen or disease-causing organism between different hosts

zero-grazing intensive method of rearing livestock that keeps them confined; feed is provided rather than letting them roam freely

ACKNOWLEDGEMENTS

This book is inspired by and dedicated to the many young naturalists, students and communities that I have had the pleasure of training and working with in East Africa.

I would like to thank Christine Kabuye, Kamal Ibrahim, Maria Vorontsova and Paul Peterson for their valuable insights, discussions and outstanding work on grasses in general. I am very grateful for the support from Struik Nature, and especially Pippa Parker, for taking on this project and publishing this work on East African grasses.

This content has been developed through collaboration with and support from many institutions over the years, including: Nature Kenya; the East African Herbarium, National Museums of Kenya; the Turkana Basin Institute; the African Butterfly Research Institute; the Museum of Comparative Zoology, Harvard University; the Kenya Horticultural Society; the Mpala Research Centre; the Kenya Agricultural Research Institute (KARI); the Kenya Wildlife Service; the East African Wildlife Society; the Royal Botanic Gardens, Kew; Stony Brook University; Princeton University; the Smithsonian Institution; the Suyian Trust; National Geographic Society; and the Food and Agriculture Organisation of the United Nations (FAO).

Information, insights, support and collaboration in many different forms have been provided by the following people: A Powys, the late G Powys, L Coverdale, D Roberts, N Pierce, D Haig, A Pringle, the late F Jenkins, W Tong, J Kingdon, L Snook, D Estes, EO Wilson, K Horton, R Leakey, M Leakey, L Leakey, G Domberger, E Whitley, A Whitley, I Angelei, P Kahumbu, S Kahumbu, B Gemmill-Herren, C Batello, N Azzu, H Herren, S Miller, T Kuklenski-Miller, R Copeland, W Kinuthia, L Njoroge, C Hemp, I Gordon, M Kasina, G Mwachala, S Masinde, P Omondi, M Kinnaird, J Mamlin, SE Mamlin, N Croze, the late E Krystall, V Otieno, P Matiku, F Ng'weno, Q Luke, SC Collins, P Lomosingo, B Obanda, T Achevi, W Okeka, J Ilondanga, C Zook, J Sandhu, PD Paterson, I Ngiru, R Plowes, L Gilbert, A Rhodes and many, many others.

I would also like to thank the staff, students and scientists of the Mpala Research Centre where I have been based for much of the writing of this book.

Mountain Anthoxanthum
(*Anthoxanthum nivale*)

INDEX TO SCIENTIFIC NAMES

INDEX TO ENGLISH COMMON NAMES